D0882765

MINING THE BORDERLANDS

Mining and Society Series
Eric Nystrom, Arizona State University
Series Editor

Our world is a mined world, as the bumper sticker phrase "If it isn't grown, it has to be mined" reminds us. Attempting to understand the material basis of our modern culture requires an understanding of those materials in their raw state and the human effort needed to wrest them from the earth and transform them into goods. Mining thus stands at the center of important historical and contemporary questions about labor, environment, race, culture, and technology, which makes it a fruitful point from which to pursue meaningful inquiry at scales from local to global.

Books in the series examine the effects of mining on society in the broadest sense. The series covers all forms of mining in all places and times, building from existing strengths in mining in the American West to encompass comparative, transnational, and international topics. By not limiting its geographic scope to a single region or product, the series helps scholars forge connections between mining practices and individual sites, moving toward broader analyses of the global mining industry in its full historical and global context.

Seeing Underground: Maps, Models, and Mining Engineering in America
Eric C. Nystrom

Historical Archaeology in the Cortez Mining District: Under the Nevada Giant
Erich Obermayr and Robert W. McQueen

The City That Ate Itself: Butte, Montana and Its Expanding Berkeley Pit
Brian James Leech

Mining the Borderlands: Industry, Capital, and the Emergence of Engineers in the Southwest Territories, 1855–1910
Sarah E. M. Grossman

MINING THE BORDERLANDS

Industry, Capital, and the Emergence of Engineers in the Southwest Territories, 1855–1910

SARAH E. M. GROSSMAN

UNIVERSITY OF NEVADA PRESS *Reno & Las Vegas*

University of Nevada Press | Reno, Nevada 89557 USA
www.unpress.nevada.edu

Jacket photographs: (*background*) courtesy of the University of Wyoming, American Heritage Center, Samuel H. Knight Papers; (*foreground*) Courtesy of The Huntington Library, San Marino, California.
Cover design by Rebecca Lown

LIBRARY OF CONGRESS CATALOGING-IN-PUBLICATION DATA
Names: Grossman, Sarah E. M., author.
Title: Mining the borderlands : industry, capital, and the emergence of engineers in the Southwest Territories, 1855–1910 / Sarah E. M. Grossman.
Description: First edition. | Reno, Nevada : University of Nevada Press, [2018] | Series: Mining and society series | Includes bibliographical references and index.
Identifiers: LCCN 2018011418 (print) | LCCN 2018016598 (e-book) | ISBN 978-1-943859-83-2 (cloth : alk. paper) | ISBN 978-1-943859-84-9 (e-book)
Subjects: LCSH: Mineral industries—Economic aspects—Mexican-American Border Region—History. | Mining engineers—Mexican-American Border Region—History. | Mineral industries—Mexican-American Border Region—History. | Mining corporations—West (U.S.)—History. | Mines and mineral resources—West (U.S.)—History.
Classification: LCC HD9506.U63 (e-book) | LCC HD9506.U63 M49 (print) | DDC 338.20979/09034—dc23
LC record available at https://lccn.loc.gov/2018011418

The paper used in this book meets the requirements of American National Standard for Information Sciences—Permanence of Paper for Printed Library Materials, ANSI/NISO Z39.48-1992 (R2002).

FIRST PRINTING

Manufactured in the United States of America

CONTENTS

ACKNOWLEDGMENTS

As all writers eventually discover, the process of producing a book is essentially collaborative. In imagining, researching, writing, and completing this book I am indebted to many friends and colleagues for their generosity, thoughtful criticism, and friendly suggestions. I've spent a large percentage of my professional life in Ithaca, New York, and I owe a large debt of gratitude to Ray Craib, for steering me out west, and to Ron Kline. At Cornell, I managed to find not one, but two exceptionally collegial corners of the university, and have some outstanding colleagues at Cornell University Press and SEAP who have supported me, whether they knew it or not, through the writing of this book: Abby Cohn, Fred Conner, Thak Chaloemtiarana, Thamora Fishel, Chiara Formichi, Roger Haydon, Kitty Liu, Tamara Loos, Michael McGandy, Kaja McGowan, Andy Mertha, Ange Romeo-Hall, and Eric Tagliacozzo.

I could never have completed the initial phase of research for this project without generous research grants from the American Historical Association, the Arizona Historical Society, the American Heritage Center at the University of Wyoming, Harvard University, the Huntington Library, and the University of New Mexico Department of History. As everyone who has had the privilege of studying there knows, the UNM Department of History has a remarkably convivial scholarly culture. It goes without saying that I would never have written this book if I hadn't landed at UNM. To the extent that I have anything useful or interesting to say, it is entirely due to what I learned there. In particular, Sam Truett is to be praised for his willingness to read draft after draft of my work and to comment constructively on each one. (As an editor myself now I can *truly* appreciate what a great skill this is!) Fellow historians, friends of long standing, professors, and colleagues who have helped me out, read my work, let me sleep on their couches and spare beds, and poked and prodded me out of research-and-writing isolation include: Eric Ash, Matt Babcock, Durwood Ball, Carin Berkowitz, Kent Blansett, Judy Bieber, Erin Brown, Cathleen Cahill, Kristen Case, B. Erin Cole,

Ariadna Forray, Meg Frisbee, Alison Games, Linda Hall, Paul Hammer, Rob Harper, Steve Hindle, Erin Hunter, Liz Hutchison, Paul Hutton, Terry Kelly, Brian Leech, Joe Lenti, Erik Loomis, Katie McIntire, Rupali Mishra, Brandon Morgan, the late Timothy Moy, Elaine Nelson, Claire Nettleton, Lindsay O'Neill, Julie Orlemanski, Sandra Rebok, Andrew Sandoval-Strausz, Jason Scott Smith, Colin Snider, Stephanie Sobelle, Katie Sosnoff, Isaac Stephens, Jason Strykowski, Brandi Townsend, Blair Woodard, and Abigail Zitin. Series editor Eric Nystrom is a generous and cheerful colleague; I count myself lucky to have met him. Working with Eric and the University of Nevada Press has been a great experience all around. Thanks are due to editor Justin Race and meticulous assistant Kit Snyder, who have done their level best to make me a compliant author. Copy editor Jeff Grathwohl's excellent work improved the readability of this book immensely, and I am very grateful.

Finally, I am fortunate in my long-suffering family, who never once suggested that studying history was an insane proposition: my parents, Bill and Gail Grossman; my brothers, Matthew and James; my best reader, Park Doing; and my littlest reader, Ms. Alice. Thank you.

MINING THE BORDERLANDS

INTRODUCTION

Industrial Transnationalism in the Late Nineteenth Century

"Are you still a candidate for the superintendency of a mine?" James D. Hague inquired of his colleague, Ellsworth Daggett. The mine in question was actually a group of silver mines in Chihuahua, Mexico, owned by the Cusihuiriachic Mining Company, an American company with offices in Boston. Hague was a consulting mining engineer; Daggett, who was Hague's choice to be the on-site superintendent of the Cusihuiriachic mines, was also an American mining engineer. Daggett accepted Hague's offer and in 1885 moved to Mexico to oversee the work of getting the mines up and running.

Previous management had made some minor improvements to the site. There was a forty-stamp mill and a lixiviation plant for the separation of silver-lead ores, but Hague thought they needed significant technological intervention to begin paying dividends.[1] He offered Daggett $500 a month, explaining, "The whole thing needs clever management, and the mine especially wants to be taken in hand by an experienced man."[2]

Hague and Daggett exchanged hundreds of letters over the next two years while working for the Cusihuiriachic Mining Company. They discussed all aspects of the mining company's operations, including the number of letters Daggett was required to produce for the investors in Boston describing how work proceeded in Chihuahua; U.S. tariff laws and how they affected the decision to ship sulfides to New York rather than bullion; the possibility of revolutionary violence among the residents of Cusi; the overall debt burden of the company; the weather; and the health and activities enjoyed by Daggett and his wife, who joined him in Mexico for a time. There were few aspects of the operation of the mines that did not fall under the purview of either Daggett or Hague, and the men took a keen interest in the fortunes of the company. Daggett kept Hague apprised of action on the ground, while Hague apprised Daggett of the mood of the investors. But their correspondence regarding the Cusihuiriachic Mining Company ended shortly after Hague announced to Daggett that the death of one Mr. Barney, an investor in the mine, "removes a principal supporter" of the operation. Hague could "foresee trouble if it has got to be supported by money drawn from here [Boston]."[3] Finding that operations in Chihuahua were not self-sufficient, both Hague and Daggett moved on to other mining ventures.

The Cusihuiriachic Mining Company was part of a broader trend of American investment in mining opportunities around the western U.S.-Mexico border, following the Mexican-American War, in the present-day states of New Mexico, Arizona, Sonora, and Chihuahua.[4] In addition to being located far from major metropolitan centers, and often far from transportation hubs or even roads or rails, these mining operations also frequently shared investors, technical experts, and even work crews. While some Americans who sought wealth in these newly publicized mineral districts were independent prospectors, the majority were involved in mining and speculation companies, both small and large. The officers and founding members of these companies were often men who had traveled through the region, either with the army or independently, but the trustees and investors tended to be less experienced either in mining or with the region. As a result, investors sought the advice of expert consultants and knowledgeable men such as Hague and Daggett—mining engineers whom they could employ in the field.

In the transnational space of the late nineteenth- and early twentieth-century borderlands, mining engineers became the critical agents responsible for drawing private capital into a region distant from the financial hubs of New York and San Francisco and relatively inaccessible by road or rail, even by the standards of the times.[5] Prior to making a mining deal, speculators, financiers, and company bosses relied on mining engineers for accurate surveys and valuations of selected ore bodies. After purchase, mining engineers drew up operating plans; mapped the underground works; selected and installed machinery; and, if they were talented and/or lucky, did all of this within a budget that ensured a return on investment.[6] As mining engineers were situated at the nexus of technical expertise and cultural influence, they had tremendous influence on how private American capital was invested during an era notable for the rapid expansion of the U.S. economy and the westward movement of large numbers of people, citizen and immigrant alike.

The correspondence between James Hague and Ellsworth Daggett concerning the Cusihuiriachic Mining Company illustrates two of the most common types of work undertaken by mining engineers in the late nineteenth century: field management (Daggett) and consultation (Hague). Their roles speak to their relative prominence within the profession. Hague was among the earliest generation of German-educated American mining engineers to seek a career in the mining fields of the western U.S. and northern Mexico. He was also a financier and investor in his own right and went on to phenomenal success running a large mine at Grass Valley in northern California.[7] Daggett, a few years younger than Hague, received his training at the Sheffield Scientific School, the future engineering college of Yale University, and worked during his career in numerous mines throughout Mexico and the southwestern U.S.[8]

In a general sense, James Hague worked at Cusi as an overseer, and he engaged in lengthy correspondence with the principal investors in the property as the chief technical advisor. He gained his knowledge from his own initial survey of the property, plus extensive communication with Daggett. Daggett, armed with information from Hague regarding the company's financial situation, the attitude of the investors, and budget guidance, had a relatively free hand in day-to-day operations, determining which mines to focus on, assessing technological needs, and overseeing personnel. Between them, Hague and Daggett made virtually all the significant decisions at the mine, from choosing the site of initial investment (Hague), to devising a plan of operations (Hague and Daggett), to overseeing labor, choosing equipment, and bringing it in (Daggett), and determining where and how profits could be made (Hague and Daggett). This pattern of shared responsibility for mine decisions and operations was repeated at mine sites throughout the borderlands, and mining engineers wielded unequaled influence as they directed the expansion of capital into the region.

Western Mining and Its Discontents

During the late nineteenth and early twentieth centuries, mining and mine engineering were completely transformed. Changes in the North American mining industry were driven by two major forces: the consolidation of U.S. political and military power across the continent, and heightened demand for the base metals needed to support the country's industrialization. In particular, the market for copper expanded substantially during the latter decades of the nineteenth century due to the growth of the new electrical industry. Demand for the metal easily surpassed available supply from the well-established copper mines in the Lake Superior region.[9] In turn, mineral speculators sought out copper resources in the new western territories. The most successful of these initial ventures were in Montana and the southwestern borderlands.

Copper deposits in these territories differed significantly from the majority of contemporary Lake Superior ores. Although there were some major pockets of high-grade western ore, most of the newer copper deposits were widely dispersed and comparatively low-grade. Mining engineers were crucial to the project of learning how to extract and exploit such deposits economically. Because of the particular constraints of mining in the borderlands, including the expense and difficulty of transporting ore to market and the relatively complex composition of the copper ore bodies, many of the key technical developments in working highly dispersed sulfides and porphyry ores, such as sulfuric acid leaching, were formulated by mining engineers working there.[10]

By the twentieth century, professional mining engineers had become critical to success in the mining industry. Yet although the copper market was a transformational force in American mining, it was not the only factor in the rise to prominence of professional mine engineering. Rather, the confluence of the general growth of the mining industry with the increasing professionalization of the white-collar technical and scientific work force radically altered the organization of mining labor. The history of mining in the trans-Mississippi West was heavily influenced by the patterns of mine exploration, discovery, and labor that were established during the California gold rush. The boom in California mining in 1849, and subsequent rushes to (for instance) the Comstock Lode (1859), western Montana (1863), and Deadwood (1872), were driven by the discovery of precious metals in riverbeds and streams, accessed by placer mining for the gold or silver that had washed from a mountainside. After these placers "played out," more industrial mining methods were employed—usually either hydraulic mining (hydraulicking) or hard-rock underground mining. These required a comparatively larger investment in capital, labor, and management than placers.[11]

Placer mining, with pan, sieve, and sluice, required great patience but little capital. The best way to learn was through experience. Seasoned veterans understood which rock formations were most likely to harbor the usually elusive motherlode—the body of ore feeding the placer deposits (hopefully)—and they staked claims and set up pans and sluices accordingly. Such readily exploited placers, which could legitimately be worked by a single person or a small team or company of independent miners, usually ran out quickly, setting the stage for the more labor- and capital-intensive work of underground mining. The attempt to find and follow a vein of ore across underground faults, folds, and cracks was a different undertaking altogether, requiring a more complex skill set as well as greater investment in labor.

Where an enterprising miner could learn the basics of placer mining in an afternoon and pick up cues from more experienced miners over the course of a few months, underground lode mining required a more comprehensive understanding of geology and significantly more brute force. Inexperienced miners floundered for months learning to move rocks efficiently and how to read a landscape, often with catastrophic results. Anything more complicated than a simple hole in the ground required a basic understanding of mechanics: how to brace an underground tunnel against cave-ins, for instance; how to ventilate a mine; or how to safely remove water from the bottom of a mine shaft.[12]

Because of the complicated nature of even the most basic underground metals mining, the workforce at most western mining projects quickly became hierarchical, with less room for small-time entrepreneurship than is suggested by the enduringly romantic figure of the lone prospector seeking fortune in the wilderness. By and large, miners were wage laborers. Many worked a twelve-hour shift overseen by a foreman or superintendent, an experienced miner with some seniority and usually a rudimentary education. Others were contract workers who worked in teams that hired themselves out to a foreman for a set fee and guaranteed quantity of production. Contract work was the most common form of labor in the coal mines of Cornwall, in Britain, and was well established in North America in regions that attracted a high number of experienced Cornish workers and their descendants.

Regardless of whether miners engaged in shift work or contract work, however, they, like placer miners, learned their skills on the job. Foremen and superintendents were apprentice-trained workers and did not go to school to learn how to mine.[13] The preeminence of underground lode mining in western North America, however, exposed the weakness of a mining industry heavily reliant on apprentice-trained labor. Although plentiful, western ores tended to be difficult to work and expensive to manipulate. In short, there was space in the western mining industry for educated mine workers who had studied metallurgy, chemistry, and the physical and geological aspects of mining and who were trained to take a holistic view of a mining enterprise. Enter the mining engineers.

The Industrial Economy of the Late Nineteenth Century

The closing decades of the nineteenth century saw a tremendous influx of people from Asia and Europe into the states and territories of North America, as unprecedented numbers of workers from China, Ireland, Italy, and eastern Europe poured into cities such as San Francisco, New York, and Philadelphia. Although many worked for several years in the U.S. before returning home, others remained, bringing their families with them and changing the demographic character of the country.[14] The availability of inexpensive land for American citizens was also a characteristic of the young nation, and the ease with which citizens were able to move around the country, especially into the West, set it apart from the European nations that had formerly claimed colonial dominance over the continent. Private citizens and the federal government agreed that the bounty of the continent was theirs by divine right. Despite the resonance of the phrase "Manifest Destiny," the U.S. government had to work quite hard to gain political and military control over its western territories, exemplified by the drawn-out process of fixing the border with Mexico after the annexation of the Southwest in the 1840s, and by the brutality of federal Indian policy in the 1870s and 1880s.[15] Many private citizens, both those confident in their status as Americans and those who faced discrimination, had a pragmatic interest in gaining western land and personal wealth.

The federal government was deeply invested in the work of expansion and nation-building for economic as well as political gain. When the southwestern territories were annexed at the close of the Mexican-American War, many railway boosters looked to these territories for a southern transcontinental route to the Pacific. Indeed, this belief was behind the Gadsden Purchase of 1854 in which the United States purchased a large swath of land in present-day Arizona south of Tucson. The western territories, with their vast stands of timber and presumed rich mineral deposits, promised resources for exploitation.[16] The industrialization of the U.S. economy coincided with an era in which the nation's populace was on the move, and the steady stream of workers heading west starting in the 1850s proved immensely valuable to the capitalists and financiers who sought to provide entrepreneurs with the raw materials needed to create an industrial manufacturing economy. Industrialization was accompanied by gradual mechanization, the deskilling of wage labor, and the growth of a hierarchical industrial and corporate bureaucracy.[17]

My study of mining engineers focuses geographically on those who worked in the various mining districts surrounding the U.S.–Mexico border in and below Arizona and New Mexico. Mine engineering is a transnational profession. Ore beds and rock formations are unconstrained by national boundaries, and mining follows suit. Nor were American financiers in the nineteenth century content to keep investments within the boundaries of the nation. Rather, they extended their influence across borders, creating American enclaves or using American money and expertise to build companies that exploited rich local resources with relatively inexpensive local labor.[18] Mining engineers encouraged and supported this transnational work. In part, this is because the geological coherence of the region led mining engineers to see the borderlands as a singular space, with its own characteristic ores and a unified set of logistical problems. Having worked at one area mine, a mining engineer could reasonably argue that he had solid knowledge of the local ores and thus could obtain a position at a neighboring mine. In consequence, many engineers who traveled to the borderlands as young men continued to work in the same region for much of their careers. In addition, the comparative isolation of the borderlands, as compared to mining territories in Colorado or California, made this transnational space a distinctive mining region. Such isolation, particularly in the nineteenth century, discouraged frequent travel, and so mining engineers were compelled to take advantage of their sometimes-brief stays in the territory by circulating through neighboring mining districts to study local conditions and mining practices. It was rare for a mining engineer who traveled to Sonora not to visit mines in Arizona. In turn, engineers who traveled to Chihuahua were likely to make side trips north to New Mexico or Arizona.[19]

By the 1880s, in spite of the international boundary line, political and technological developments in the United States and Mexico served to knit the region more closely together. Seven-term Mexican president Porfirio Diaz (1876–1880, 1884–1911) opened his country to international investment, and American dollars poured in, favoring north-south rail lines and Mexican mines.[20] Mining engineers passed easily and frequently across the border. During the 1880s, copper also began to be successfully mined along the southernmost edge of Arizona and the northernmost edge of Sonora. As the market for copper wire boomed with the coming of electricity in the 1890s, many financiers began to see the copper borderlands as mining engineers always did: a coherent geological landscape. This in turn increased American enthusiasm for Mexican mining ventures and heightened interest in the region in boardrooms, clubs, and investment houses in New York and San Francisco. Under the management of development and exploration companies, mining engineers also worked at both new and well-established mine sites to catalog the mineral wealth of the region and to devise plans for its exploitation. As a region focused on mineral wealth and documented by mining engineers, the western borderlands can be described as a *technocratic landscape*. Its objective value was relative to the ability of companies to extract minerals, orchestrated by mining engineers.

A regional study of mining engineers in the southwestern United States and Mexico also highlights the international reach of the moneyed interests for whom the engineers worked, while underscoring the complex work such men did mediating between field and work, and investor and mine.

Expertise and Professionalization

The quintessential expert mining engineer in nineteenth-century North America was hired by the president or board of directors of a company to report on and improve conditions in the field. He had some claim on knowledge that a mere skilled miner lacked: deep experience working ores similar to those under consideration; broad knowledge of the specifics of regional mining conditions; and significant scholarly understanding of mining problems, treatments, and technologies. Furthermore, he could communicate this knowledge both to the workers on site and, more importantly, to the managerial bureaucracy of the mine in a way that could be easily understood. By communicating clearly with a board of directors about field conditions, an engineer demonstrated that he understood the mining industry and that he was sensitive to the relationship between cost of production and profit margin. Similarly, a successful mining engineer was able to discuss mining techniques and ore appearance with experienced miners and foremen, demonstrating that he understood the technical skill necessary to work an ore seam or lode and could guide the miners in a successful enterprise, rather than forcing them to labor for a mine that would not pay.

Expertise is difficult to define. It has a tendency to appear to be primarily a consensus on the part of relevant constituents to grant scientific and/or technical authority to a class of privileged, well-connected "experts," but it also entails possessing specialized, technical knowledge. In the case of mining, the rise to prominence of mining engineers resulted in substantive changes in labor and market relations throughout the industry. Surely both aspects of expertise—the *possession* of specialized knowledge, as well as the *consensus among elites* that mining engineers knew something other workers did not—drove these changes.[21] In the western borderlands, mining

Cross-section of an underground mine, from an introductory mining textbook, depicting two mining engineers discussing how best to follow the "true fissure" vein to the surface. Etienne Ritter, *From Prospect to Mine* (Denver: Mining Sciences Publishing Company, 1910), 88. Courtesy of The Huntington Library, San Marino, California.

engineers tended to play down their class privilege and play up their technical knowledge. There were many within the western mining industry, however, who were deeply suspicious of the knowledge claims of mining engineers and saw them as university-trained ninnies—Easterners brought in to tell the locals what to do.[22] The balance between the class status of mining engineers and their rhetorical strategies for gaining and maintaining authority, along with the benefits accrued by their backers stemming from their specialized technical knowledge, was critical to their success.

Recognition that mining was a complex industry that could benefit from the guidance of expert professionals did not occur in a vacuum. In other sectors of society, doctors and lawyers sought to institutionalize their work, and beginning in the 1840s began establishing professional gatekeeping societies and advocating for more formal university training. Engineering, a profession that gained a foothold in the United States with the construction of the Erie Canal in the 1830s, followed suit, albeit slowly and in decentralized fashion, as mechanical engineers began to organize in the mid- to late-nineteenth century. Mine engineering established formal patterns of training and professionalization several years after civil and mechanical engineering began such institutionalization, likely because there were large numbers of apprentice-trained miners who found the time to read extensively in trade magazines and publish books about the science and technology of mineral extraction. Further, the need for the external certification of university degrees was less apparent than in other fields of engineering. In the mid-nineteenth century, many such self-taught men used the title "mining engineer" or described themselves as "scientific miners" without the credential of a formal education. This was particularly true of mining engineers who reached the apex of their careers in the 1870s and 1880s. In the 1850s, there were only a small number of formally-trained mining engineers in North America, virtually all of whom were trained at European academies, such as the Royal School of Mines in London, École de Mines in Paris, and the Freiberg Academy in Saxony. By the 1890s, mine engineering was sufficiently established as a course of study that a man who wanted a position as a mining engineer needed several semesters in a mining course, even if he never completed the full E.M. degree (Master of Engineering, as it was awarded at the time).

In the 1870s and 1880s, however, the population of mining engineers working in the west had a more varied educational background. For instance, one of the most important mining engineers in the borderlands, James Douglas of Phelps Dodge, did not have a formal degree in mine engineering. A chemist by training, Douglas taught at Morrin College in Quebec for several years before pursuing a career as a mining engineer. By the time he did his first really significant work for Phelps Dodge, in 1880, he had been managing mines and performing consultations for over a decade. Douglas was always aware that he had a less specialized background than many of the engineers who were his close associates, but it is worth noting that his actual education probably did not differ too much from that of mining engineers who attended Freiberg in the 1860s. The establishment of an entrenched class of specialized mining engineers is a critical feature of the development of the mining industry in the late nineteenth century (discussed in detail in chapter 2), but prior to 1900 the most pertinent distinction between specialized mining engineers and other mine workers was not the existence of a formal mining degree; it was class-enabled scientific and technical literacy.

The breadth of activities in which mining engineers engaged was generally recognized by the nascent engineering industry. When the first engineering societies were founded in the United States, the American Institute of Mining Engineers (AIME) was remarkable for being fairly open in its membership.[23] Mining engineers were happy to welcome capitalists and financiers to their conferences, possibly because it was helpful if their employers better understood their work. More likely the AIME's policies were an acknowledgment that mining engineers had to be closely attuned to the desires of their financiers. Mine engineering requires money. Therefore, mining engineers believed that money men should be intimately involved in their business.

The ability to tell wealthy financiers and mine owners that their investments were not going to pay out was a key aspect of their work, especially in the early years of the profession. In addition to hotly discussed codes of ethics, mining engineers relied on an important self-identified characteristic of their professional identity: their institutionalized and somewhat romantic belief that they were self-reliant, practical men whose job was to speak plainly and directly at all times. Throughout the late nineteenth and early twentieth century, many aspects of mine engineering changed, including the credentialing process, the job trajectory of an individual engineer, the content of university training, and many, if not most, features of the work itself. The professional identity of mining engineers, however, remained remarkably constant over more than a half-century of significant industrial change and the physical and economic transformation of the borderlands.[24] As they were captured into corporate bureaucracies, they continued to see themselves as rugged, independent individualists.

Why the Southwest?

This study focuses on the experiences of mining engineers in the southwestern borderlands, rather than the breadth of metals mining in the West, in part because the region can readily act as a microcosm of North American metals mining in the nineteenth century. All the major activities occurred along the border in a region that was culturally as well as geographically well-defined: placer gold mining; silver boom; industrialized copper. To follow such trends in the broader West, one would have to travel to the California gold fields; to Deadwood and the Black Hills; across the Sierras to the Comstock Lode; down to Pike's Peak and Leadville; and over to Montana. The relatively smaller scale of most of the projects in the border region, compared to, for example, the creation of the instant cities of Virginia City or Deadwood, makes it easier to tease out the work of engineers from that of other players within the mining enterprise.

Smaller local differences also make the Southwest a useful study. Engineers working in Butte in the nineteenth century, for instance, were heavily engaged in litigation over extralateral rights. Of course, this sort of legal wrangling also occurred in Arizona but to a somewhat lesser degree (although not in Mexico—a significant distinction). Thus I have chosen to focus on the work of engineers in the field rather than in the courtroom.

Engineers and the Borderlands: A Preview

Between 1881 and 1901, copper production in the Arizona Territory expanded from an annual output of 10,000 to 1.3 million tons.[25] The work of mining engineers explicitly supported the interests of capital, generally personified by investor groups and boards of directors. They also directly oversaw workers, had a much better idea of labor conditions at the mine than other members of the management hierarchy, and were responsible for the purchase of industrial equipment and training workers to run it. By the twentieth century, mining engineers at the larger mining companies along the border—including Cananea Consolidated Copper Company, Phelps Dodge, Arizona Copper, and Calumet and Arizona—were among the chief proponents, and in some cases the orchestrators, of the segregationist and paternalistic policies that became important features of towns dominated by the mining industry such as Bisbee or Morenci in Arizona, or Cananea or Nacozari in Sonora.[26] That mining engineers, whose training and expertise equipped them to be technical consultants, were ultimately responsible for making decisions in all the varied aspects of mine management, is crucial to understanding the development of the mining industry through the borderlands.

For engineers in the U.S. Southwest and northern Mexico, the work of mediating between capital and labor was animated by factors specific to the region. A central concern was logistical. It was more difficult to transport equipment into, and bullion out of, southern Arizona or Chihuahua in the 1860s and 1870s than in Colorado or Nevada. The distance from rail lines and the expense of building rails into the desert, combined with the scarcity

of water and fuel to power pumps, mills, and smelters and the challenge of recruiting American workers to such distant (and frequently climatically inhospitable) locales, created almost insurmountable difficulties that slowed or challenged the development of many otherwise promising mining opportunities. Of course, other mining districts could also be hard to reach. But there was a difference of scale between the logistical complications of the more populated mining districts of the Rockies and those of the border. In Colorado, for example, a "delayed" development of an area was a couple of years. Leadville, considered a weak opportunity in 1876, was Colorado's second largest city in 1878. Further, there were many other opportunities for mining engineers and workers in the region. Once the boom was on, workers and experts were on hand to solve the immediate problems. In the borderlands, a "delay" could be a decade. There weren't smelter works in the next gulch because there might not be *people* in the next gulch.[27]

For mining engineers working in Mexico, another concern was tariff law. Even if mining could be profitably carried on in Mexico, and bullion could be shipped out with relatively little expense, would all profits be lost in taxes at the U.S. border? If so, was the cost of shipping bullion to London or Paris worthwhile? This was an issue for Daggett and Hague in their work at the Cusihuiriachic Mining Company. Although they ultimately abandoned that project for other reasons, U.S. tariffs were of major concern to the two engineers.

An additional and occasionally overarching borderland problem was that of culture. Some engineers viewed the opportunity to work in Mexico or the southwestern territories of the United States as a tremendous adventure. Such men embraced the foreignness of their surroundings and were tolerant of, or intrigued by, the linguistic and cultural divisions between themselves and the average mine worker. The ability to speak Spanish was obviously a crucial dividing line between mining engineers who loved the borderlands and those who loathed it. Language skills were rarely a job prerequisite, and many engineers were happy to learn enough Spanish on the job to communicate effectively with their foremen and local dignitaries. Others chose to presume there was no need to learn anything of local society and permitted their racial and cultural biases to color their interactions with mine workers.[28] Not all workers at these mines were Mexican or Mexican-American, of course. Native American labor was also consistently used through the nineteenth century, particularly in Mexico. By the turn of the century, Arizona also had large populations of European immigrant miners. Towns such as Bisbee were known as "white-man's camps" in which only white Anglo or European workers were hired for the more lucrative underground positions.[29] Mexicans, Mexican-Americans, and Chinese employees of mining companies stayed above ground shoveling ore, doing laundry, and cooking. The racial and ethnic diversity of the workforce was a defining characteristic of working life in the mining districts surrounding the U.S.–Mexico border.

An additional feature of the region was the migratory nature of labor. There was tremendous turnover among underground workers in mining camps throughout the area.[30] Mining engineers themselves were also extremely transient. Almost by definition, their work was peripatetic, taking them from one company to another and one mining site to another on a regular basis. Engineers readily crossed state, territorial, and national lines, pursuing the most interesting or most lucrative work they could find.

Another relatively consistent attribute of their work throughout the nineteenth and early twentieth centuries was that mining engineers functioned as on-the-ground representatives of the business interests of mining companies. Fieldwork was the great draw of mine engineering, but it could also be uncomfortable and occasionally dangerous. In most cases, mines were far from company headquarters, and mining engineers were themselves unprotected from the consequences of edicts issued by the main office—a significant issue as industrialization in mining led to increasingly antagonistic relations between miners and management. Some scholars have accused mining engineers of wielding technical expertise as a weapon in the war of management against labor. While this interpretation goes too far in attributing malevolent intentions to mining engineers, it is important to acknowledge that on the whole they were not overly concerned with the working and living conditions of laborers. Indeed, it is striking how infrequently mining engineers refer to labor unrest or to union organizing, except during the violent labor upheavals of the early twentieth century.[31]

In the mid-nineteenth century, as American interest in borderlands mining began, mining engineers were primarily hired as surveyors. Working alone, or with one or two assistants, they were dispatched to likely locations by investors. There they surveyed surface and subterranean ore bodies, sketching rough maps and mining plans for their employers. Depending on the financial situation of the mine owner or investors, the engineer might remain on site as a technical expert or as the mine manager. Over the next several decades, this general pattern shifted. In part, this was due to the increasing importance of copper in the economies of the border states and territories.

Copper's influence was not restricted to the border region, of course. The first and most significant location to benefit from the demands of the emerging electrical industry was the Anaconda mine in Butte, Montana, where "the richest hill on earth" generated well over nine million pounds of copper per season beginning in 1882, after the railway reached Butte and copper ore could then be moved east economically.[32] The southwestern mining territories remained harder to reach, but when copper demand outstripped even the production of the Anaconda, the potential for profit brought industry to the borderlands.

The effect was profound in the Arizona Territory. From 1880 onwards, copper accounted for approximately 20 percent of territorial economic output. By 1900, Arizona was responsible for 50 percent of U.S. copper production. The neighboring Mexican state of Sonora also had several large copper mines in locations such as Cananea and Nacozari.[33]

The emphasis on copper also shaped mining practices in the border region, in part because of the high demand, but also because of the nature of the copper deposits. Unlike the pure native ore found near Lake Superior, copper deposits in the Rockies and Southwest were of much lower grade. Keeping up with market demand using these low-grade ores required new mining practices. A miner could not simply follow a vein of ore if there was no vein to follow. Rather, copper ore and its rock matrix had to be removed from the ground together and then treated extensively before being shipped to market. The systems devised by mining engineers to accomplish this task changed the structure of working life in mining camps, in effect turning borderlands mining into an industry based on what historian Timothy LeCain has labeled "mass destruction"—the use of heavy technology and unskilled labor to dismantle a landscape on a massive scale for the production of metals.[34] These new techniques also had a profound effect on the working lives of mining engineers themselves.[35]

The chapters that follow are organized in a rough chronology to delineate the tension between changes in engineering practice and the stability of certain aspects of mining engineers' professional identity over the course of more than a half century of dramatic expansion in the mining industry. Mid-nineteenth century mining engineers were members of an elite class of European-educated men from prosperous eastern backgrounds. By the early twentieth century, the establishment of systematic engineering education democratized the profession to some degree, and the growth of corporate mining eliminated much of the autonomy characteristic of mine engineering in the early decades. Despite these changes, mining engineers remained wedded to the notion that their profession rewarded its members with working lives distinguished by self-reliance, a concept increasingly at odds with the quotidian reality of mine engineering.

The work of mining engineers on the treatment and extraction of complex ore bodies was instrumental in creating and supporting a vision of the American industrial landscape as a quantifiable exploitable space—a technocratic landscape.[36] The expansion of American capital into the borderlands was thus inextricably bound up with the emergence and changing role of a new class of technical professional: the mining engineer.

Notes

1. Lixiviation is a way of precipitating silver or silver, lead, and/or copper from roasted ores and was a common leaching process in the late nineteenth century. The precipitating agent in lixiviation is hyposulfite, in which silver chloride is readily soluble. According to mining engineer Carl August Stetefeldt, lixiviation was first introduced in North America by Guido Küstel and was used extensively throughout northern Mexico and Arizona, in part because the process was cheaper than the local Mexican patio process. In lixiviation, upfront costs are significantly lower than in amalgamation. Ores need only be coarsely crushed; there is no need for an expensive purchase of mercury; and much smaller engines and boilers are required. See Carl August Stetefeldt, *The Lixiviation of Silver-Ores with Hyposulphite Solutions, with Special Reference to the Russell Process* (New York: Scientific Publishing Company, 1888), vi–vii, 3–4, 112–134.

2. James D. Hague to Ellsworth Daggett, November 23, 1885; Hague to Daggett, December 7, 1885; Hague to Daggett, January 26, 1886; all L9, James D. Hague Papers, The Huntington Library, San Marino, CA (hereafter JDH).

3. Hague to Daggett, February 9, 1886–February 19, 1887; all L9, JDH.

4. Although the Arizona Territory was not officially severed from the New Mexico Territory until 1863, I use the term "Arizona" throughout this book to refer to the area encompassed by the present-day state of Arizona. This is in keeping with the usage of locals and even of American travelers to the region, who either called the territory "Arizona," "Arizuna," or some version of "Apache-land" or "Apacheria" to distinguish it from the eastern portion of the territory the United States formally claimed as New Mexico.

5. The historiography of the extension and integration of the infrastructure of the United States in the nineteenth century is long. Critical classic works include Alfred Chandler, *The Visible Hand: The Managerial Revolution in American Business* (Cambridge: Belknap Press, 1977); Robert H. Wiebe, *The Search for Order, 1877–1920* (New York: Hill and Wang, 1967); Howard Lamar, *The Far Southwest, 1846–1912: A Territorial History*, rev. ed. (Albuquerque: University of New Mexico Press, 2000); and William G. Robbins, *Colony and Empire: The Capitalist Transformation of the American West* (Lawrence: Kansas University Press, 1994). All offer concise analyses of the way these changes

affected the southwestern United States. More recently, Rachel St. John, *Line in the Sand: A History of the Western U.S.-Mexico Border* (Princeton: Princeton University Press, 2011), and Samuel Truett, *Fugitive Landscapes: The Forgotten History of the U.S.-Mexico Borderlands* (New Haven: Yale University Press, 2006), extend the narrative by considering how these changes play out on both sides of the Mexican border.

6. Clark Spence, *Mining Engineers and the American West: The Lace-Boot Brigade* (New Haven: Yale University Press, 1970); Kathleen Ochs, "The Rise of American Mining Engineers: A Case Study of the Colorado School of Mines," *Technology and Culture* 33, no. 2 (1992):278–301; Eda Kranakis, "Social Determinants of Engineering Practice: A Comparative View of France and America in the Nineteenth Century," *Social Studies of Science* 19, no. 1 (1989):5–70; Eric C. Nystrom, *Seeing Underground: Maps, Models, and Mining Engineering in America* (Reno: University of Nevada Press, 2014); Timothy LeCain, *Mass Destruction: The Men and Giant Mines That Wired America and Scarred the Planet* (New Brunswick, NJ: Rutgers University Press, 2009); Logan Hovis and Jeremy Mouat, "Miners, Engineers, and the Transformation of Work in the Western Mining Industry, 1880–1930," *Technology and Culture* 37, no. 3 (1996):429–456.

7. Rodman Paul, ed., *A Victorian Gentlewoman in the Far West: The Reminiscences of Mary Halleck Foote* (San Marino, CA: The Huntington Library, 1972), 36, 362, 372–384.

8. "Preliminary Roll of the Sheffield Scientific School, 1846–1869," Sheffield Scientific School, Yale University, Records (RU 819). Manuscripts and Archives, Yale University Library (hereafter RU 819).

9. A general sense of the shape of the copper industry can be gleaned from copper production statistics. In 1880, Michigan still produced 80 percent of the copper produced in the U.S. But just years later that share had dropped to 38 percent, while Butte, Montana, produced 42 percent and Arizona 14 percent. By 1920, these percentages were reversed: almost half of the copper produced in the U.S. came from Arizona with a mere 13 percent from the erstwhile copper center of upper Michigan. See Hyde, *Copper for America: The United States Copper Industry from Colonial Times to the 1990s* (Tucson: University of Arizona Press, 1998), 81.

10. Rodman Paul and Elliott West, *Mining Frontiers of the Far West: 1848–1880* (Albuquerque: University of New Mexico Press, 2001), 155; Charles K. Hyde, *Copper for America*, chs. 4–5; T. A. Rickard, *The Copper Mines of Lake Superior* (New York: *Engineering and Mining Journal*, 1905); Larry Lankton, *Cradle to Grave: Life, Work, and Death at the Lake Superior Copper Mines* (New York: Oxford University Press, 1991), 24.

11. Kent A. Curtis, *Gambling on Ore: The Nature of Metal Mining in the United States, 1860–1910* (Boulder: University Press of Colorado, 2013); Andrew Isenberg, *Mining California: An Ecological History* (New York: Hill and Wang, 2005); Otis E. Young Jr., *Western Mining: An Informal Account of Precious-Metals Prospecting, Placering, Lode Mining, and Milling on the American Frontier from Spanish Times to 1893* (Norman: University of Oklahoma Press, 1970).

12. Curtis, *Gambling on Ore*, 17–20, 72–73; Kathryn Morse, *The Nature of Gold: An Environmental History of the Klondike Gold Rush* (Seattle: University of Washington Press, 2003), 92–93; 101–102; Duane Smith, *Mining America: The Industry and the Environment* (Boulder: University Press of Colorado, 1994), 5, 14; Mark Wyman, *Hard Rock Epic: Western Miners and the Industrial Revolution*, 1860–1910 (Berkeley: University of California Press, 1987), 9–14; Young, *Western Mining*, esp. ch. 7.

13. Lankton, *Cradle to Grave*, 52–53; 62–69.

14. John Bodnar, *The Transplanted: A History of Immigrants in Urban America* (Bloomington: University of Indiana Press, 1985), xv–xvi; 2–4; 57; Rebecca Edwards, *New Spirits: Americans in the Gilded Age, 1865–1905* (New York: Oxford University Press, 2005); David Gutierrez, *Walls and Mirrors: Mexican Americans, Mexican Immigrants, and the Politics of Ethnicity* (Berkeley: University of California Press, 1995).

15. St. John, *Line in the Sand*, ch. 1, ch 2.; Lamar, *The Far Southwest*; Patricia Nelson Limerick, *The Legacy of Conquest: The Unbroken Past of the American West* (New York: Norton, 1987), 18–30; Paula Rebert, *La Gran Línea: Mapping the United States–Mexico Boundary, 1849–1857* (Austin: University of Texas Press, 2001); Truett, *Fugitive Landscapes*, 13–29. American claims to exceptionalism can be so readily caricatured and demolished that it is also easy to ignore the very real differences between the nascent United States and more established nations with regard to the mobility of average citizens. Frederick Jackson Turner, of course, drew on this fact in developing his frontier thesis, and the rhetorical power of this formulation of American expansion resonated throughout the twentieth century. Alternate models of U.S. expansion were put forth by scholars such as William Appleman Williams and Walter LaFeber in the 1960s and 1970s.

With the New Western History in the 1980s and 1990s, scholars of the nineteenth century territorial expansion used the tools of post-colonial studies, ethnic history, and social history to undermine Turner's narrative. More recently, environmental historians have shifted the focus back to the remarkable ecological wealth of North America, which, perhaps inevitably, once again highlights the distinction between the early United States and Europe. A small sampling of relevant work in environmental, political, and diplomatic history includes Ned Blackhawk, *Violence over the Land: Indians and Empires in the Early American West* (Cambridge, MA: Harvard University Press, 2008); William Cronon, *Changes in the Land: Indians, Colonists, and the Ecology of New England* (New York: Hill and Wang, 1983); Karl Jacoby, *Shadows at Dawn: An Apache Massacre and the Violence of History* (New York: Penguin, 2009); Walter LaFeber, *A New Empire: An Interpretation of American Expansion, 1860–1898*, new ed. (Ithaca: Cornell University Press, 1998); Ted Steinberg, *Down to Earth: Nature's Role in American History* (New York: Oxford University Press, 2002); Elliott West, *The Contested Plains: Indians, Goldseekers, and the Rush to Colorado* (Lawrence: University Press of Kansas, 1998); Richard White, *The Middle Ground: Indians, Empires, and Republics in the Great Lakes Region, 1650–1815* (New York: Cambridge University Press, 1991); and William Appleman Williams, *The Tragedy of American Diplomacy* (New York: Norton, 1972).

16. William Cronon, *Nature's Metropolis: Chicago and the Great West* (New York: Norton, 1992), ch. 4; Alan Trachtenberg, *The Incorporation of America: Culture and Society in the Gilded Age* (New York: Hill and Wang, 1982), 7–9; 95–105; 258–270; Robbins, *Colony and Empire*, ch. 1.

17. Trachtenberg, *Incorporation of America*, 4–7; Henry Adams, *The Autobiography of Henry Adams*, reprint (New York: Vintage, 1990); Chandler, *Visible Hand*; Leo Marx, *The Machine in the Garden: Technology and the Pastoral Ideal in America* (New York: Oxford University Press, 1964); Wiebe, *Search for Order*. Richard White suggests that the railroad industrialists did not actually succeed in organizing American society as

Chandler and Wiebe asserted. Rather, railroad administrators aspired to create and implement order. Their failures and dysfunction (inefficiencies, train crashes, etc.), White argues, were a symptom of their modernity. Richard White, *Railroaded: The Transcontinentals and the Making of Modern America* (New York: Norton, 2011).

18. Truett, *Fugitive Landscapes*, 4–6; 84–99; 108–115.

19. Louis Janin Diary, 1863, HM 64294, Papers of Louis Janin (Addenda), The Huntington Library, San Marino, CA; Raphael Pumpelly, *Across America and Asia: Notes of a Five Years' Journey Around the World, and of Residence in Arizona, Japan, and China*, 2nd ed., revised (New York: Leypoldt and Holt, 1870), 38–44; D. W. Meinig, *The Shaping of America: A Geographical Perspective on 500 Years of History*, vol. 3, *Transcontinental America, 1850–1915* (New Haven: Yale University Press, 1998), 152–157.

20. Sarah Deutsch, *No Separate Refuge: Culture, Class, and Gender on an Anglo-Hispanic Frontier in the American Southwest, 1880–1940* (New York: Oxford University Press, 1987); Daniel Lewis, *Iron Horse Imperialism: The Southern Pacific of Mexico, 1880–1951* (Tucson: University of Arizona Press, 2007).

21. For an excellent breakdown of some of the key features of expertise, see Eric H. Ash, "Introduction: Expertise and the Early Modern State," *Osiris*, 2nd series, 25 (2010):4–10. Some recent scholarship concerning expertise includes: Eric H. Ash, *Power, Knowledge, and Expertise in Elizabethan England* (Baltimore: Johns Hopkins University Press, 2004); Park Doing, *Velvet Revolution at the Synchrotron: Biology, Physics, and Change in Science* (Cambridge: MIT Press, 2009); Brian Martin, ed., *Confronting the Experts* (Albany: SUNY Press 1993).

22. Lankton, *Cradle to Grave*, 74; Spence, *Mining Engineers*, 70, 76.

23. Edwin T. Layton Jr., *The Revolt of the Engineers: Social Responsibility and the American Engineering Profession* (Cleveland: Case Western Reserve University Press, 1971), 29.

24. As Ruth Oldenziel demonstrates, linking engineering to traits considered masculine in the nineteenth century, such as common-sense and physical coordination, was a key component of the professionalization of the field as a whole. Ruth Oldenziel, *Making Technology Masculine: Men, Women, and Modern Machines in America, 1870–1945* (Amsterdam, NL: Amsterdam University Press, 1999), 10–12. Where mine engineering departs from that norm is in its emphasis on the romance of mining. Mining engineers wholeheartedly embraced the belief that working on a mine was an adventure, rather than a job, which colored their subject-position within the nascent corporatization of the mining industry to a remarkable extent. On changes to the industrial system of the United States, see David Hounshell, *From the American System to Mass Production: The Development of Manufacturing Technology in the United States* (Baltimore: Johns Hopkins University Press, 1984), and David Noble, *America by Design: Science, Technology, and the Rise of Corporate Capitalism* (New York: Knopf, 1977).

25. *Historical Statistics of the United States*, table DB73-78, Millennial Edition Online.

26. On company towns, and the impact of southwestern mining companies on towns that were not formally planned "company towns," see Katherine Benton-Cohen, *Borderline Americans: Racial Division and Labor War in the Arizona Borderlands* (Cambridge, MA: Harvard University Press, 2009), 106–108; John S. Garner, ed., *The Company Town: Architecture and Society in the Early Industrial Age* (New York: Oxford University Press, 1992); Linda Gordon, *The Great Arizona Orphan Abduction* (Cambridge, MA: Harvard University Press, 1999), 172–185; Monica Perales, *Smeltertown: Making and Remembering a Southwest Border Community* (Chapel Hill: University of North Carolina

Press, 2010); Duane Smith, *Rocky Mountain Mining Camps: The Urban Frontier* (Lincoln: University of Nebraska Press, 1973); Truett, *Fugitive Landscapes*, 106–108.

27. Duane A. Smith, *The Trail of Gold and Silver: Mining in Colorado, 1859–2009* (Boulder: University Press of Colorado, 2009), 102–104.

28. Jonathan C. Brown, "Foreign and Native-Born Workers in Porfirian Mexico," *American Historical Review* 98, no. 3 (1993):798–801; Joseph F. Park, "The History of Mexican Labor in Arizona During the Territorial Period" (Master's thesis, University of Arizona, 1961), 40.

29. A. Yvette Huginnie, *'Strikitos!': Race, Class, and Work in the Arizona Copper Industry, 1870–1920* (PhD diss., Yale University, 1991), 44–60; Benton-Cohen, *Borderline Americans*, 81–85.

30. Juan Gomez-Quiñones, *Mexican American Labor, 1790–1990* (Albuquerque: University of Mexico Press, 1994), 44; Deutsch, *No Separate Refuge*, 18–20, 88–90.

31. A. Yvette Hugginie, "A New Hero Comes to Town: The Anglo Mining Engineer and 'Mexican Labor' as Contested Terrain in Southeastern Arizona, 1880–1920," *New Mexico Historical Review* 69, no. 4 (1994): 323–344. On mining labor history, see Thomas Andrews, *Killing for Coal: America's Deadliest Labor War* (Cambridge, MA: Harvard University Press, 2010); Ronald Brown, *Hard Rock Miners: The Intermountain West, 1860–1920* (College Station: Texas A&M University Press, 1979); James Byrkit, *Forging the Copper Collar: Arizona's Labor-Management War of 1901–1921* (Tucson: University of Arizona Press, 1982); Elizabeth Jameson, *All that Glitters: Class, Conflict, and Community in Cripple Creek* (Urbana: University of Illinois Press, 1998); Richard Lingenfelter, *The Hardrock Miners: A History of the Mining Labor Movement in the American West, 1863–1893* (Berkeley: University of California Press, 1974); Wyman, *Hard Rock Epic*. Of great interest is the Bisbee deportation file, box 181, John and Isabella Greenway Collection, MS 181, Arizona Historical Society, Tucson, Arizona (hereafter Greenway Collection).

32. Richard E. Lingenfelter, *Bonanzas and Borrascas: Copper Kings and Stock Frenzies, 1885–1918*, vol. 2 (Norman: University of Oklahoma Press, 2012).

33. *Historical Statistics*, Table DB73-78; Hyde, *Copper for America*, 127; Marvin Bernstein, *The Mexican Mining Industry, 1890–1950* (Albany: SUNY Press, 1967), 72–73; David T. Day, *Mineral Resources of the United States [1902]* (Washington, D.C.: Government Printing Office, 1904), 164–166.

34. LeCain, *Mass Destruction*, 7–11.

35. Many histories of mining labor focus on technological change either implicitly or explicitly as the trigger for the violent industrial actions of the IWW and the WFM of the 1890s and early 1900s. Scholarship that explicitly considers the role of technological change in labor relations includes Brown, *Hard Rock Miners*, ch. 5; Huginnie, "'Strikitos!'"; Larry D. Lankton and Jack K. Martin, "Technological Advance, Organizational Structure, and Underground Fatalities in the Upper Michigan Copper Mines, 1860–1929," *Technology and Culture* 28, no. 1 (1989):43–45; Mouat and Hovis, "Mining Engineers"; Wyman, *Hard Rock Epic*, esp. ch. 4. Studies of the mining labor force that emphasize social and political issues over technological change include Benton-Cohen, *Borderline Americans*; David R. Berman, *Radicalism in the Mountain West 1890–1920: Socialists, Populists, Miners, and Wobblies* (Boulder: University Press of Colorado, 2007); Jameson, *All That Glitters*; and Lingenfelter, *Hard Rock Miners*.

36. John G. Gunnell, "The Technocratic Image and the Theory of Technocracy," *Technology and Culture* 23, no. 3 (1982):392–416.

CHAPTER ONE

EARLY MINING IN THE BORDERLANDS

The Limits of "Intelligence and Capital"

"The simple truth is, that the soil of nearly all North America is more or less impregnated with gold.... In Arizona and Sonora it is known to abound. In Mexico, it would have been a principal article of export but for the greater plenty of silver."[1] This hyperbolic statement from *Harper's Weekly* in 1858 was perfectly in step with what was commonly printed about the rich mineral resources of western North America in the years following the California gold rush. Reports of the fabulous wealth unearthed there were commonplace to readers of papers in New York or Chicago. Thus, when reports of other sites of mining wealth began to circulate, whether the Comstock Lode in 1859 or Pike's Peak several years later, the American public was primed to accept them, and speculators and entrepreneurs prepared to take action to reap the benefits of the continent's natural wealth. American fortune seekers who traveled to Nevada or Colorado did so in company with thousands of fellow countrymen.

But some Americans took a less well-traveled path in the 1850s and 1860s—down to the newly mapped U.S.-Mexico border region. The Arizona and New Mexico Territories, and the states of Sonora and Chihuahua in Mexico, were each known to have several mining districts. Independent prospectors, speculators, and members of joint-stock companies hastened south to capitalize on the promise of the region. The vast majority of these entrepreneurs were not particularly successful, but their experiences were of tremendous importance to those who followed. Early prospectors popularized the region as a destination for adventurers and provided information about what might be needed to establish successful local mining operations.

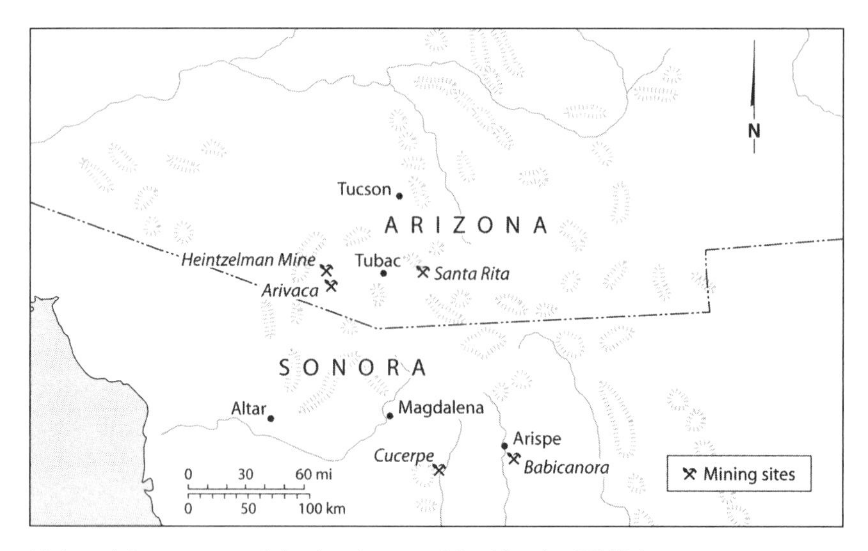

Major mining centers of the border, ca. 1860. Map by Bill Nelson.

Participants in this initial influx were a disparate group. Some came west from the more populous states. Others, particularly in the early 1850s, traveled southeast from the gold-mining districts of northern California. For many of these erstwhile forty-niners, the gold rush was an experience of thwarted dreams. The expense and difficulty of reaching California meant that by the time would-be miners arrived in San Francisco they had very little money and even less interest in the tedium of panning for gold. Lacking immediate success, many who still had the means to leave did so, including Americans, Mexicans, and other foreign nationals. The latter left in particularly large numbers following passage of the Foreign Miners' Tax in 1850. This charge of twenty dollars per month for non-Anglo miners to run a placer operation led many gold seekers from Sonora, in particular, to head home from the California mines. Some were sidetracked by silver mines in the New Mexico Territory, but many more headed back into Sonora or Chihuahua, embracing the opportunity to return to Mexico while trying their luck prospecting in established mining districts near Hermosillo, Alamos, or Batopilas.

American fortune hunters joined the Sonoran natives, drawn by tales of Mexican mineral wealth, the promise of easy travel, and the economic integration of Sonora and Arizona.[2] Not all miners in the gold and silver districts of Sonora, Arizona, and Chihuahua came via California. Contemporary reports counted upwards of a thousand men who left Alamos and Hermosillo, in Sonora, for the territory of the Gadsden Purchase.[3] Anticipating its utility for a transcontinental railroad, the United States acquired this strip of land across southern Arizona in 1854 from Mexico. That there was gold and silver in the mountains near Tubac and Tucson was an unexpected bonus for American legislators.

Upon reaching the mining districts in the borderlands, prospectors found a series of overlapping ethnic, national, and political groups operating throughout the region. Although the most recent major political disruptions in the area were the annexation of the New Mexico Territory and the Gadsden Purchase by the United States, the relative absence of the U.S. military in the aftermath of the Treaty of Guadalupe Hidalgo meant that relations between the new American territories and the northern Mexican states were fairly low key. Indeed, the border between the United States and Mexico was almost completely permeable and scarcely marked through the Sonoran and Chihuahuan Deserts. Law-abiding citizens of both countries crossed with ease, while bandits and Native Americans exploited the diplomatic agreements that prevented soldiers and authorities on either side of the border from giving chase into the other country. Ease of border crossing meant that American prospectors were as likely to head south of the border as to stay to the north. Whether in the United States or northern Mexico, prospectors found that federal authorities had little to do with local affairs—a condition greatly exacerbated in U.S. territory with the advent of the Civil War in 1860. Instead, they were at the mercy of more local concerns.

Among other inhabitants, several native tribes called the borderlands home. A number of them maintained generally peaceable relations with Sonoran and American authorities. These included the Akimel O'odham (known to contemporary Americans as the Pima), who occupied the Gila River basin north of Tucson; the Tohono O'odham (known to locals as the Papago), who traveled through a broad swath of land on either side of the Santa Cruz River, encompassing the old presidios of Tucson, Tubac, and Tumacacori and crossing the new international boundary line; and, farther south, the Yaqui. The presence of these and other native tribes did not prevent Americans from often referring to the area along the border between present-day Arizona and Sonora as Apachería or Apache-land, in reference to the people with the strongest military presence in the region. Most Mexican and American citizens were ignorant of the existence of multiple Apache tribes, and, indeed, many white travelers to the borderlands made a habit of calling any native person who made them slightly nervous "Apache."[4] Travelers' reports from Arizona and Sonora, in particular, focus on the military threat of the Apache, and to the extent that the United States maintained a military presence in and around Tucson in the 1850s, it was to combat Apache claims on the newly American territory.

Local political conditions in mining camps in the borderlands region were further complicated by endemic civil unrest throughout the state of Sonora in the 1850s and 1860s. Occasional Apache raids were only the start of the problem for Sonoran miners. The actions of American filibusters—private U.S. citizens who marched mercenary armies into the state in a bid to claim it for the United States via conquest—made native Sonorans cautious of the *yanqui* newcomers. Most prospectors who settled into mining communities near the new international boundary line, however, had little interest in the actions of the filibusters except as they disrupted work on mining projects.[5] In addition to invasions by American soldiers of fortune, during the 1850s and 1860s Sonora also endured a more successful foreign occupation, this one by French forces. As the French had some interest in the supposed wealth of the silver mines of Sonora, the state experienced an extended occupation. The governor of Sonora, meanwhile, fled north into Arizona, while French forces, allied with the Yaqui and Mayo tribes, assaulted the population. This uneasy state of affairs did not settle down until the late 1860s, when a weakened French army was finally pushed out of Sonora.[6]

In the midst of all this political and military turmoil, prospectors and miners still flocked into a region whose mines were legendary. In the tradition of California mining, many were independent prospectors. During the California gold rush, placer mining was the norm. A miner separated flakes of gold or silver in a wide pan from stream sediment washed down from a higher location (the elusive "mother lode"). Placer mining requires almost no capital outlay other than a pan and a sieve. One man working alone is as likely to find a nugget of value as a group of men working with a marginally more complex trough or sluicing system. In the arid Southwest, however, there were few active placer mines in the 1850s.[7]

Many of the larger ore deposits in the borderlands were silver rather than gold. Much of this silver ore was embedded in quartz rather than carried into streams, and so the riverine mining experience of many former California gold seekers did not help much once they reached the mining districts of Chihuahua, Sonora, or present-day Arizona. Those newcomers who mastered different mining techniques, however, were vocal in their endorsement of the local mineral wealth. One mining engineer, pleased by the encouraging assays his silver samples from Batopilas received in San Francisco, remarked, "The more I have become acquainted with the California mining, the better I am satisfied with the value of the Mexican."[8] Abundant gold and silver was manifest in the region in narrow veins of almost pure ore, making it relatively simple to extract with pick-axe and shovel.

Although Americans used words such as "inexpert," "minor," and "desultory" to describe ongoing mining projects that they observed, there was a great deal of active mining. Mexican practices followed methods the Spanish brought in the sixteenth century. While mining was not the largest industry in the borderlands, it occupied an important place in the local economy, with many people working occasional claims or hiring out as skilled labor at the larger mines. Such mines rarely used the wood timbering practices that were common at underground mines in the United States and Europe. Rather, ore was left in situ as pillars to prevent roof collapse. Smelting and processing techniques first used by the Spanish and still widely practiced were labor- but not capital-intensive and enabled small-time producers to work with some small success even refractory ores (those that manifested in concert with other minerals rather than as pure veins). If a prospector hit a good strike, local knowledge of how to build an *arrastra* or implement the *patio* method was readily available and inexpensive, if a person were so inclined.[9] There were ample experienced Mexican and "friendly" natives, members of the Papago, Ópata, and Yaqui tribes, who could be hired to work in a promising mine.[10] Some Americans, however, were uninterested in learning from local experience. As newcomer John Denton Hall fumed, "It was impossible...to give satisfaction" to such men. "Many a poor devil of a Mexican miner, on giving a correct assay...and report of a mine, has been belied and abused as an ignoramus...from the simple fact that he has not satisfied by lying, the hope of his much more ignorant employer."[11]

Sonora

The career of John Denton Hall exemplifies in many ways the experiences of the new settlers in the border territories. An Englishman who met with "indifferent success" in the California gold fields, Hall accepted the invitation of a Mexican friend he met there to try his luck mining silver in Sonora.[12] For the next fifteen years, he engaged in a series of unsuccessful mining operations, both in Sonora and southern Arizona. Hall's memoir, *Travels and Adventures in Sonora*, offers a notably detailed depiction of the experience of mining silver in the borderlands in the mid-nineteenth century for a certain kind of white foreigner. Like many Americans who shifted from California to Sonora, Hall had no connections to New York investment money and was constantly struggling to raise capital. But unlike many other white men who traveled to the region, Hall became remarkably attached to Sonora during his residency, experiencing the borderlands as an interesting, if not always comfortable, place to live, rather than simply an exotic background for adventure. In short, he liked it there.[13]

Hall was loquacious on the subject of Sonora's mineral wealth. Indeed, he announced an intention in the preface of his memoir to make it clear that the mines of Sonora and Arizona were "safe investments for legitimate mining," a truth Hall feared was lost on the American public due to the machinations of "unprincipled speculators." Unfortunately for his stated purpose, the narrative that follows is more notable as a chronicle of indifferent success. Hall detailed his own failures and those of other investors as well. In the spring of 1852, for instance, he began working at a gold mine at El Cajon de Brisca. By June 5th, 1852, he related, "we were to all intents and purposes dead broke." Later that summer, Hall partnered with another American, Henry Clark, and the two men did some small-time "gold digging" nearby.[14] Reading between the lines, it is evident that what the two men did was camp out and live hand to mouth from what game they could shoot. Tiring of this life, Hall and Clark began working with a Mexican-owned silver mining company at the Santa Teresa de Jesus mines, just north of the mineral district of Cucurpe, and about fifty miles from the U.S.-Mexico border south of Tucson. "We were dubious as to the success of the speculation," Hall explained, "but there was something so enticing about silver mining, that we accepted the offers made [to] us."[15] Indeed, Hall returned to gold mining only two more times in the next fifteen years, for ventures in which, needless to say, he did not turn a profit.

Following the failure of the Santa Teresa mining effort, Hall and Clark again decided to strike out on their own, independently working a mine close to Cucurpe. When this enterprise also failed, Hall apprenticed himself to the new amalgamator at the Santa Teresa mines, continuing to work in the district until 1855, when the company stopped work. Hall next contracted to work the recently abandoned mines with some workers supplied by the company. The attempt was cut short by a wave of violence, this time the civil war in Sonora. Hall's workers were drafted to fight for the government.[16] Frustrated by the disruption of his speculations, Hall shifted his attention northward to Arizona, where he entered into an unwritten contract with a man who claimed ties to investment money in

New York. This project, too, failed miserably, and Hall returned to mining in Sonora. Once again, he was disappointed. As he put it, "Hopes apparently so well-founded have all vanished like smoke before the wind."[17]

In *Travels and Adventures in Sonora*, Hall described seventeen different mining operations in Sonora and Arizona, active between 1850 and 1866, which were owned whole or in part by foreign investors, the vast majority of whom were American. During this period, many Americans were enticed south of the border by the loose legal strictures on mining and the willingness of eastern investors to be persuaded that Mexican mines held greater potential than the proven wealth of California mines. One contemporary estimated that in the year 1860–1861, as many as twelve fully financed American companies began mining operations in Sonora alone.[18] Of those operations in Sonora or Arizona with which Hall himself was familiar, only three resulted in anything but abject failure: the Babicanora mines, operated after 1864 by a French corporation; Sylvester Mowry's Patagonia mine in Arizona, at which operations were suspended at the start of the American Civil War; and Hall's own small success with a mine located near the Santa Teresa, circa 1863.[19]

Although John Denton Hall was a small-time operator, and his travails were probably unknown to people outside his immediate circle, the arc of his career in Sonora and Arizona, from independent prospector, to prospecting with other people's money, to working for larger mining companies, is emblematic of the general trend in the borderlands hard-rock mining industry before the Civil War. The directors and investors in regional mining companies were acutely aware of the difficulties posed by work in a region so distant from the economic center of American society. Yet they and other observers preferred to blame their failures on aspects of their industry that were more mutable than location. Sylvester Mowry, for instance, blamed the workers, stating baldly, "The Mexican is poor, without energy, and too lazy to trust to help himself."[20] Hall, who fancied himself a mining expert (which Mowry was not), saw a different problem. He believed that "the great difficulty in working mines in Sonora and Arizona to advantage, is the scarcity of scientific, practical miners; for this reason, parties are diffident of investing in mines, fearful of losing both mine and money."[21]

Credentialed mining experts agreed with Hall's assessment of the situation in Sonora. Louis Janin, for instance, a German-trained engineer, visited Arizona and Sonora twice between 1862 and 1864. His second journey was paid for by none other than the Santa Teresa Mining Company with which Hall had been involved (by 1863 the mine was owned by a syndicate based in San Fran-

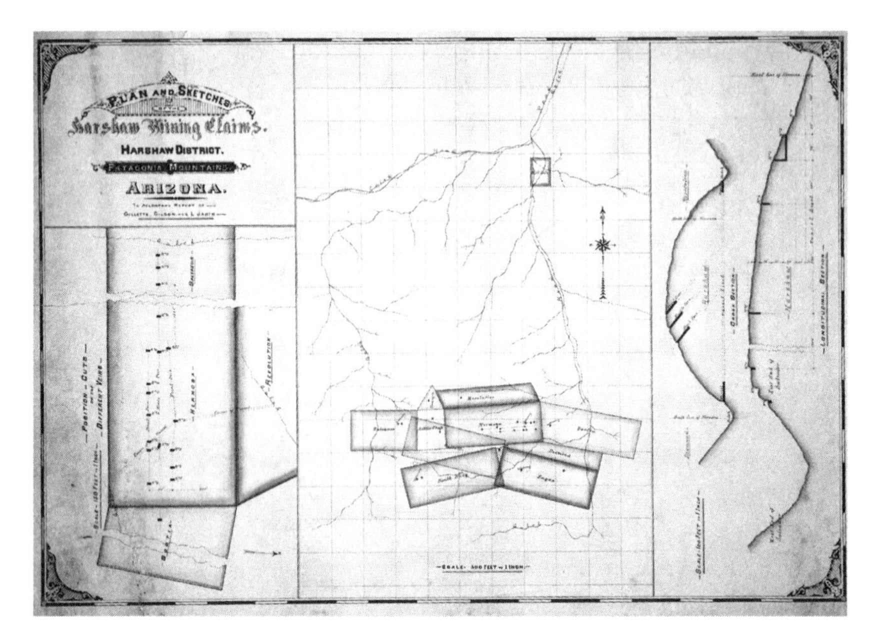

This map, drawn by Louis Janin during a trip to the border in the 1880s, exemplifies how mining engineers mapped claims and districts. Louis Janin, "Plan and Sketches of the Harshaw Mining Claimes, Patagonia Mountains, Arizona, c. 1880," Charles Janin Collection box 45 (1). Courtesy of The Huntington Library, San Marino, California.

cisco). In his formal report to the company, Janin argued, "With a competent mining engineer at their head, the yield [at the Santa Teresa] could be largely increased within a few months."[22] Privately, he took all the mining enterprises he visited around the borderlands very seriously. Pleased with the quality of the ore he found, Janin outlined extensive suggestions in his notebooks regarding the necessary technologies and processes to improve mining efficiency. But despite his obvious enthusiasm for the possibilities of the region, Janin also noted that implementing technical improvements was challenging. At the small El Mortero mine in northern Sonora, for instance, he found that almost none of the new equipment that the mine had recently purchased was installed. "The engine + boiler was in place," he noted, but "not even one barrel frame [is] in order...[and] no furnaces are up."[23] Because of such problems, Janin was a proponent of the patio process and the relatively easily implemented barrel amalgamation (Freiberg method) for separating silver (discussed below).

Although Hall's own story was one of failure, he clearly hoped that his experiences would attract investors who had knowledge of mining and more financial resources. As Louis Janin's observations indicate, Hall's analysis of the borderlands mining industry was in sync with the opinions of the formal mining establishment in the United States. Indeed, such opinions took on the status of gospel truth, as even the *New-York Daily Times* declared, "There are very rich silver mines in this country [Mexico and New Mexico], but neither capital, enterprise, nor knowledge, to work them."[24]

The Arizona Territory

Assembling both knowledge and capital at a mine site in the borderlands proved quite difficult in the 1850s and 1860s. Hall, for instance, was largely an autodidact. But once he learned how to work the regional ores, he was never able to find real success with the small amounts of capital to which he had access. A handful of other operators, however, were able to raise fairly substantial sums of cash and brought into the area bona-fide mining professionals—engineers—as consultants and managers of various mining operations. Many of these larger businesses were founded as joint-stock companies, overseen, sometimes loosely and sometimes tightly, by boards of trustees or investors east of the Mississippi. Beginning in the 1830s, industries requiring massive up-front investment, such as railroads, and later, mines, were almost exclusively financed in the United States by sales of stock. That John Hall did not have the connections, and therefore the access, to these sources was a factor in his inability to capitalize on the opportunities he had in Sonora.

The advantage of the joint-stock system for company managers was the relative ease with which large sums of money could be raised. Using stock sales to finance borderland mines also brought the mineral wealth of the territory to the attention of financiers, whose investment through the late nineteenth century was necessary. Mining entrepreneurs became railroad boosters who needed cash to build regional infrastructure.[25] Although stock was often sold in a manner similar to today, in a public offering, it was also sold piecemeal. For instance, the owners of the Gila Copper Mine decided in 1857 to sell a few shares in their company for cash to ship already-extracted ore out of Arizona.[26] Of course, such ease in raising capital came with the price of nominal accountability to shareholders for the success or failure of a company's operation.[27]

With stockholders' money, mining companies hoped to purchase the latest mining and smelting technology and to hire the experienced workers who could operate such equipment. Advertising the skill and credentials of staff mining engineers thus became a crucial tool for managers and owners seeking to raise money.

The Heintzelman Mine

The Sonora Exploring and Mining Company, overseen by a board of trustees in Cincinnati, was one of the most well-financed mining companies in the borderlands region in the 1850s. Initially capitalized at over $2 million, a remarkably large sum at the time, it had all the advantages that observers thought a successful borderlands mining company needed: money, a property with substantial silver ore holdings, and some of the top mining engineers working in North America.[28] Yet the inability of even this well-advised and well-financed company to pay dividends indicates the hollowness of the standard analysis of mining, as expressed by Hall and others, that success simply required money and knowledge. One contemporary mining journal put it clearly: "[Mining] requires of those who pursue it a special education and experience, or the labor devoted to it may be fruitless...it is the offspring of intelligence and capital."[29] Yet, as the history of the Sonora Exploring and Mining Company demonstrates, this formulation, while logical, is not complete.

The Sonora Exploring and Mining Company was founded by Charles Poston and his business partner, Major Samuel Peter Heintzelman. Poston was a regional booster who lobbied Congress to separate the Arizona Territory from New Mexico. Heintzelman, a veteran of the U.S.–Mexico War, had stayed in the region after the Treaty of Guadalupe Hidalgo, establishing an Army post at the junction of the Gila and Colorado Rivers in 1850.[30] Like many military men stationed on the frontier, Heintzelman spent some of his spare time surveying the landscape for speculative mining prospects. In 1856, he and Poston purchased a silver claim near Tubac in the Gadsden Purchase territory, eponymously named the Heintzelman Mine, and the two established a company to exploit the claim, located in the Santa Rita Mountains just north of the Mexican border.

When raising capital for their venture, Heintzelman and Poston placed great emphasis on the *security* of the Sonora Exploring and Mining Company's property, assuring investors that the "mere presence" of four companies of dragoons in the nearby town of Tucson, and, of course, Heintzelman himself, would prevent Apache raids at the mine.[31] Indeed, the first annual report, likely written by Poston himself, told stockholders that the company had sent "an armed party [to the Gadsden Purchase] of sufficient strength to protect itself against the Indian tribes."[32] This statement follows a trope of eastern American writing about mining in Mexico or the new southwestern territories in which the armaments carried by the exploring party were described in as much detail as the supposed value of the mines.[33] The Sonora Exploring and Mining Company aggressively asserted the security of its property, no doubt hoping that this would direct stockholder's attention to the subject of most interest to all: mining.

Poston and Heintzelman's strategy in this regard proved moderately successful. But from the start they were plagued by other difficulties. The company had trouble getting the necessary credit from freighting companies in Sonora to transport expensive and heavy machinery from the nearest ports in western Sonora. This meant that they had to try to bring it to Tubac overland from San Francisco or points in the eastern U.S. Although his Cincinnati investors appreciated the apparent security of a company managed by a major in the U.S. Army, Heintzelman proved to be a terrible executive. He was far more interested in his military career than the day-to-day life at the mine; he disliked the southern Arizona climate; and he appointed his monumentally ineffective brother-in-law, Solon H. Lathrop, as manager of the mine.[34]

Despite shortcomings in the managerial department, one area in which the Sonora Exploring and Mining Company distinguished itself was the hiring of mining engineers, the majority of whom had trained at Freiberg, in Saxony. Unusually for the region, the Heintzelman Mine often had on staff more than one mining engineer, and as a contemporary observer noted, the company made an effort to hire a "real class of people."[35] This is a testament to the relative solvency of the operation, which gave the Sonora Exploring and Mining Company the ability to hire expensive experts whose skills could be touted in annual reports and trade publications. One of the company's earliest public announcements of its plans stated that annual production was projected to be almost half a million dollars, a "view... borne out by the agent of the Company and *scientific gentlemen* on the expedition [my emphasis]."[36]

In the 1850s and 1860s, mining engineers working in the border territories, or in Mexican states, were only rarely hired to manage mining properties. The Sonora Exploring and Mining Company followed this pattern. Over a seven-year period, 1856–1863, the company employed mining engineers Guido Küstel, Charles Schuchard, Frederick Brunckow, and Herman Ehrenberg as metallurgists, prospectors, on-site engineers, and technical advisors. None served as a supervisor or manager, although all were central to the company's ability to advertise its operations. Küstel, for instance, wrote books on ore dressing and processing gold and silver ores and eventually became one of the best-known engineers in the western United States. He came to Arizona from a smelting firm in San Francisco, lured by the richness of the ore in the

Santa Ritas.[37] His reports on the value of the Heintzelman mine were cited by mining professionals and regional boosters alike.[38] Hiring Küstel sent a clear message to investors that the company was able to attract educated mining men to its ores and served to proclaim the seriousness of the endeavor to anyone who might question it.

Mining engineer Herman Ehrenberg was also noted by company directors to stockholders as a particularly valuable asset. In addition to writing articles in mining journals, Ehrenberg was a frequent contributor on mining matters in the local paper, the *Weekly Arizonian*, and the paper returned the favor by praising his talents to the sky. Although far from an objective reporter of matters of local interest—the *Weekly Arizonian* was owned and operated by the Santa Rita Mining and Milling Company, a subsidiary of the Sonora Exploring and Mining Company—the high opinion of Ehrenberg expressed in the editorial pages was seconded by interested easterners, one of whom described him as "a gentleman of education and intelligence and thoroughly informed upon all points connected with these distant countries [Arizona and Sonora]."[39]

Yet despite the wealth of intellectual capital at the Heintzelman Mine, the mine struggled to turn a profit. Guido Küstel reported in 1857 that he was able to extract $148 of silver per ton of ore. While not an insignificant payout, this was hardly enough to cover operational costs. When it opened, the Heintzelman used an adobe oven to heat ore prior to treatment and treated the ore using the patio process, a traditional Mexican system for separating silver ore from quartz. The oven was a simple blast oven, operated by a single man working a bellows.[40] Although a supremely effective technology for a small-scale local mining operation, the patio process could not produce enough silver to pay the cost of shipping to markets in New York or San Francisco. To be profitable, the Heintzelman needed to operate on a more ambitious scale, producing relatively large quantities of high-quality silver. The company invested in mining engineers to enable profitability. Although mining engineers were comparatively expensive employees, the amount of money that could be made by selling high-grade silver was expected to be far more than the cost of one or two engineers.

Guido Küstel initially took responsibility for upgrading the Sonora Exploring and Mining Company's treatment processes. Shortly after beginning operations, he instituted the barrel process of amalgamation for on-site reduction of ore. Processing silver ore—separating the silver from the mined matrix and forming it for easier transport—is complicated. As Küstel explained, "The concentration of silver ores is generally a delicate process, being subject to heavy loss, which cannot be avoided."[41] Barrel amalgamation, also known as the "Freiberg method" after the German institute where it was developed, was a process for treating silver ores by mixing them with salt, using silver's affinity for the ionized chlorine in salt to separate it from what nineteenth-century mining manuals called "baser materials." Half a ton of roasted, pulverized ore was placed in a barrel, along with a large quantity of water and iron. The barrel was then rotated at a high speed on a vertical axis. This continued until the "paste" in the barrel reached a consistency similar to heavy cream. At that point, mercury (quicksilver) was added to the barrel, and the rotating continued for another twelve hours, after which the barrel was topped off with water and the rotation slowed so that an amalgam fell to the bottom of the barrel. As a final step, the amalgam was reheated to remove the mercury, and the remaining silver concentrate shaped into blocks for transport.[42]

Initially, Küstel used a team of mules to power the rotation of the barrels. This was not very successful. The trade publication *Mining Magazine* reported that the mules provided insufficient "regularity of motion" for successful amalgamation. In addition, there was "much injury and loss occasioned by stopping to rest or change the animals."[43] The engineers at the Heintzelman wanted to bring in steam-powered equipment to solve these production issues, but the nearest port, Guaymas, was fifteen hundred miles away, and the directors feared the route was not secure. The other option was to ship the amalgamator overland in at least two distinct pieces. Each piece of equipment weighed 50,000 pounds, and bringing them overland would cost the company $15,000.[44] Despite the impracticability of this cost for a mine producing approximately $500 per week, management decided to buy the equipment, calculating that the long-term viability of the mine required such investment.

Yet despite all the calculated risks the various mining engineers suggested to Heintzelman, Lathrop, and their successor, industrialist Samuel Colt, the Heintzelman Mine failed to live up to its promise. Indeed, the engineers, rather than increasing the output of the mine, seem to have served instead as convenient scapegoats for more general managerial problems. Heintzelman, for instance, publicly reprimanded one of his mining engineers for lacking knowledge of metallurgy. He further accused the engineer of wasting money through lenient dealings with Mexican laborers and of not understanding barrel amalgamation, a process about which Heintzelman himself assuredly knew almost nothing.[45] Given Heintzelman's general ignorance of mining processes, it is difficult to give much credence to the major's attacks on his mining engineers, especially as Heintzelman's final, ill-conceived action as company president was to send a shipment of silver overland to San Francisco for smelting, where from "some imperfection in the furnaces, or some other cause, [the yield] was 47 per cent. less than the assay. This scarcely paid the expense of sending the ore so far and was the source of much embarrassment to the Company."[46] Such poor decision making was not unusual at the Sonora Exploring and Mining Company, where valuable shipments of equipment were perpetually getting lost or stalled in a tangle of bad business practice and personal vendetta.

Yet the mining engineers who worked for the company were prolific publicists of their work, issuing a multitude of articles devoted to their metallurgical practices and the geology of Arizona in trade publications such as New York's *Mining Magazine* and in local, company-owned newspapers such as the *Weekly Arizonian*. Positive reports on the wealth of the mine continued to be issued through the 1850s, although such reports were almost always accompanied by an enumeration of the logistical difficulties in bringing modern mining and metallurgical practice to the Heintzelman Mine.[47] The Sonora Exploring and Mining Company lacked neither investment capital nor intellectual capital in the form of mining experts. Rather, it suffered rather from poor management and the perhaps-foreseeable difficulty of trying to mine a quartz district several hundred miles from the nearest water or rail route. In addition, bad luck and bad timing plagued the enterprise, as the federal government withdrew troops from Arizona at the start of the Civil War in 1861, and investors concurrently pulled out of mining enterprises for the duration of the conflict.

The Santa Rita Mining and Milling Company

Neighboring the Heintzelman Mine was a subsidiary of the Sonora Exploring and Mining Company, financed by the same group of Cincinnati-based capitalists: the Santa Rita Mining and Milling Company. Like the Sonora Exploring and Mining Company, the Santa Rita Company made an effort to hire university-trained mining engineers, although without the same capital resources, it only succeeded in hiring one, a young man named Raphael Pumpelly, in 1860. Like Guido Küstel, Pumpelly was a graduate of the Freiberg Academy.[48] Still, like the Sonora Exploring and Mining Company, the Santa Rita Mining and Milling Company possessed both "intelligence"—in the form of Pumpelly—and "capital."

For a company operating in the multicultural environment of the borderlands, Pumpelly was a good hire. Although he did not speak Spanish, he was fluent in French, and his employers may have hoped that he would be able to muddle through the language barrier to communicate with Mexican laborers, as proved to be the case. As a recent Freiberg graduate who was also proficient in German, the language in which much of the cutting-edge research in mining methodology was published, Pumpelly was acquainted with the most up-to-date scholarship. He was also connected to a wide network of professional mining engineers who could provide the young engineer with the benefit of their experience if he so required.[49]

Yet although the directors made the calculation to hire a Freiberg-trained expert, Pumpelly did not have more success than Ehrenberg, Küstel, et al., at the Sonora Exploring and Mining Company. Pumpelly's task in Arizona was to help the Santa Rita Mining and Milling Company open some new mines and to establish a modern smelter in the Santa Rita Valley. In a region that, in Pumpelly's own words, was "credited by Mexican tradition" to possess silver deposits of unthinkable wealth, his job was to make such rumors into reality.[50] As noted by William Wrightson, the general manager of the Santa Rita Mining and Milling Company, "It is one thing to have ore—however good it may be, and quite another thing to extract the silver out of it."[51]

Almost from the start, Pumpelly's attempts to institute modern mining methods in Arizona were plagued with difficulty. When he first headed to Arizona in 1860, the Santa Rita was already deeply in debt, acquired in part from the company's sponsorship of the *Weekly Arizonian.*[52] He was not impressed by the landscape or climate. This native of Owego, New York, considered the locality of Tubac to be "a veritable hell."[53] Most problematic for Pumpelly, he realized shortly after his arrival that "the capital of our company was not proportionate to the results expected to be achieved."[54] Despite his misgivings, Pumpelly spent considerable time mapping the Santa Rita's holdings, which he considered impressive and intrinsically rich. Indeed, Pumpelly traded on his time in Arizona later in his life, giving public lectures on the ores of Arizona and publishing a memoir of his time at the Santa Rita as well as producing a couple of short travelogues for publications such as *Putnam's* in the 1860s.[55] During the first weeks of Pumpelly's time in Arizona, he surveyed the company's property, visited neighboring mining sites, and experimented with various means of heating the company furnace, which did not correspond to any furnace Pumpelly had previously encountered and for which he was the sole responsible party.[56]

Despite being one of the most credentialed young mining engineers in the United States at the time, Pumpelly was unable to raise the fortunes of the Santa Rita Mining and Milling Company. Unlike at the Heintzelman Mine, where the technical experts were publicly lauded for their skill and privately blamed for the company's failure to turn a profit, the directors of the Santa Rita Mining and Milling Company expressed more primal fears about the problems facing their business, going so far as to tell their investors that the area surrounding Tubac was a "country which is worse than a frontier... [it] is, in reality, the heart of a wilderness, with laborers who speak a different language from our own."[57] During his tenure in the borderlands, Pumpelly was by his own account preoccupied with his personal safety when he was not frustrated by the lack of resources at his disposal. He believed, possibly correctly, that the Santa Rita mine was the site of more Apache raids than any other place in the country and considered the attempt to operate a modern mining company under such conditions almost criminally insane.

Pumpelly's existential concern became particularly acute about six months after his arrival when Mr. Grosvenor, Pumpelly's friend and the superintendent at the Santa Rita mine, was killed less than a quarter mile from their shared quarters. Neither Pumpelly nor the bookkeeper, also an American, knew whether Grosvenor had been killed by Apaches in retaliation for an American raid on a local encampment the previous week or by Mexican employees in retribution for the Santa Rita Mining and Milling Company's failure to make payroll. Following Grosvenor's death, Pumpelly abandoned any pretense of working as a mining engineer and busied himself working as debt collector and accountant, attempting to extract the Santa Rita Mining and Milling Company from Arizona with as little financial loss as possible. He and his American colleagues moved to Tubac from the mine on June 15, 1861. A few months afterwards, Pumpelly left Arizona altogether, in company with Charles Poston and a man he described only as a "known murderer" named Williams.[58]

Pumpelly's experience at the Santa Rita was slightly different from that of Ehrenberg or Küstel at the Heintzelman Mine, yet it offers a similar lesson about the relative value of mining engineers in the borderlands during the 1850s and 1860s. John Denton Hall and others may have pointed to the need for capital and expertise in building a successful mine, but the publicly held companies near Tubac were also in desperate need of a more stable political climate and a cheap and safe means to transport their product to market. Pumpelly, for instance, was unable either to establish systematic extraction methods or to process successfully any quantity of ore. Yet he nonetheless proved very useful to his employers, albeit in ways they could not have anticipated. For instance, after Grosvenor's death, when it was clear that the company would be unable to continue operations, Pumpelly oversaw six weeks of smelting at the Santa Rita. During that period, he and his largely

Mexican workforce were surrounded, and occasionally shot at, by a force of Apache warriors. After the smelting was complete, Pumpelly took it upon himself to personally separate the silver from the lead *planchas*, a process that took him approximately sixty hours of nonstop labor. Since the workers at the Santa Rita had stayed with Pumpelly at risk of their own lives for several weeks, he could probably have trusted them to perform this final task without skimming too much of the profits. But given the strength of the ethnic hostility of white Americans toward Mexicans, Pumpelly's employers were undoubtedly thankful for what they would have considered his exemplary caution.[59]

Raphael Pumpelly's experience demonstrates that despite the need for technical men in the border region in the 1860s, noted by John Denton Hall among others, in the absence of military security or significant capital resources, mining engineers were not particularly valuable at this time.[60] Company directors clearly hoped that mining engineers would bring specialized knowledge into the region, elevating a mining prospect to the status of a worthy investment. The Santa Rita Mining and Milling Company, for instance, noted in its first annual report, "The unexpected difficulties which have...surrounded the Sonora Exploring and Mining Company...have created some distrust in the success of mining in Arizona." The company then noted that it sought the opinion of Frederick Brunckow of the Sonora Exploring and Mining Company, "a gentleman well known as an eminent mining engineer," regarding where it ought to sink its own mining shaft.[61] The implication is clear that whatever the troubles of the Sonora Exploring and Mining Company, the inadequacies of its technical professionals did not cause them.

In a certain sense, mining engineers during this time were most valuable to their employers as tools of the stock market, trotted out for the sake of investors and cast aside, or blamed, when profits failed to materialize. Raphael Pumpelly was hired in part to combat the distrust that eastern investors felt toward western mining ventures. Yet although the company could afford to pay for his services as a surveyor and metallurgist, the Santa Rita company could not afford to implement Pumpelly's engineering plans, nor could they afford—on a more basic level—the necessary security to protect their property or employees from death and dismemberment.

The engineers of the Sonora Exploring and Mining Company published many articles about the process of mining silver in the Santa Rita Mountains, and confidence in their abilities ultimately led one of the most important industrialists of the mid-nineteenth century, firearms manufacturer Samuel Colt, to purchase the mine. But the mine nonetheless failed to turn a profit, limited by the same dearth of local infrastructure and capital poverty as the neighboring Santa Rita mine. Company directors hoped that mining engineers would be able to solve the kinds of problems that plagued small-time operators such as Hall, and consulting engineers such as Louis Janin certainly supported this belief. But turning a profit from mining the U.S.-Mexico borderlands proved extremely difficult in the years prior to the American Civil War, and impossible once hostilities broke out in the East.

Still, the experiences of these early mining engineers were not without value. Metallurgists such as Küstel gained skill working southwestern ores and published widely read papers on their properties and attributes. Potential investors in San Francisco and New York heard stories about the great wealth of the ores mined at Alamos, Tubac, and Santa Teresa, and these stories primed potential investors in later decades. Practices were established during these years on which generations of future mining engineers continued to rely.

Lured by the challenge of working difficult ores in far-away places, mining engineers such as Küstel, Ehrenberg, and Pumpelly headed to the borderlands, each hoping he brought the necessary intelligence to work the ore and trusting that his employers would be willing to invest the necessary capital to extract it. To this ambitiously noble pursuit of excellence in knowledge, mining engineers coupled the frisson of travel to an exotic (to them), scantily mapped territory claimed and inhabited by people quite foreign to the lives of university-educated men from New York or Washington, D.C. As the profession of mine engineering grew and changed, the experiences of these early American mining engineers came to occupy a larger-than-life place within their profession. Ensuing generations of mining engineers took on that adventurous mantle for themselves.

Notes

1. *Harper's*, vol. 12/11 (1858):786.

2. Susan Lee Johnson, *Roaring Camp: The Social World of the California Gold Rush* (New York: Norton, 2000), 31; John R. Robinson Diary, 57–58, 66, HM 62476, The Huntington Library, San Marino, CA; Miguel Tinker Salas, "Sonora: The Making of a Border Society, 1880–1910," *Journal of the Southwest* 34, no. 4 (1992):435–436. The political integration of Sonora and Arizona was occasionally a topic of discussion in American politics. See for instance Amy Greenberg, *Manifest Manhood and the Antebellum American Empire* (New York: Cambridge, 1995), ch. 1; Ernest May, *The Southern Dream of a Caribbean Empire* (Baton Rouge: Louisiana University Press, 1973), esp. ch. 6; John

Douglas Pitts Fuller, *The Movement for the Acquisition of All Mexico, 1846–48* (Baltimore: Johns Hopkins University Press, 1936).

3. "Silver Mines of Arizona," *Mining Magazine* 2, no. 1 (January 1854).

4. The territorial maps provided in Karl Jacoby, *Shadows at Dawn*, provide a wonderful visual guide to the complicated political negotiations between the O'odham, the Apache, and the two national entities.

5. Rudolph F. Acuña, "Ignacio Pesquiera: Sonoran Caudillo," *Arizona and the West* 12, no. 2 (1970):148; Greenberg, *Manifest Manhood*, 31.

6. Stuart Voss, *On the Periphery of Nineteenth-Century Mexico: Sonora and Sinaloa, 1810–1877* (Tucson: University of Arizona Press, 1982).

7. In the early nineteenth century, a handful of placers were worked in the Ortiz and San Pedro Mountains, just north of Albuquerque, by Mexican miners. Productive placers were also worked near present-day Silver City by white Americans, beginning in the 1860s. Paul and West, *Mining Frontiers*, 156.

8. John R. Robinson Diary, HM 62476.

9. After being excavated, mid- and low-grade ore needed further treatment before it could be smelted. At its most basic, an *arrastra* consisted of a circular clay base, overlaid with a slightly acidic surface of stones, onto which roasted ores were dropped. A team of mules then dragged a heavy weight around the circle as a slow trickle of water dripped onto the crushing surface, mashing the ores into a slurry. The slurry dropped through the stone sieve and was retrieved from the clay base for further processing. Young, *Western Mining*, 65–70; Rossiter Raymond, *Statistics of Mines and Mining in the States and Territories*..., vol. 869 (Washington, D.C.: GPO, 1870), 3.

10. Miguel Tinker Salas, *In the Shadow of the Eagles: Sonora and the Transformation of the Border During the Porfiriato* (Berkeley: University of California Press, 1997), 11, 82–83; Paul and West, *Mining Frontiers*, 156.

11. John Denton Hall, *Travels and Adventures in Sonora: Containing a Description of Its Mining and Agricultural Resources, and Narrative of a Residence of Fifteen Years* (Chicago: J. M. W. Jones Stationery and Printing Company, 1881), 135.

12. Hall, *Travels and Adventures*, 9–10.

13. On the problems of mining in Arizona and the New Mexico Territory prior to the 1880s, see Paul and West, *Mining Frontiers*, 155, 157–158.

14. I am grateful to Sam Truett for identifying Henry Clark. Personal communication, November 18, 2013.

15. Hall, *Travels and Adventures*, preface, 93–95, 103, 109.

16. Hall's knowledge of these mines was subsequently cited as the most comprehensive in the district. Louis Janin, "Report of Louis Janin, Jr.," in *Reports, Prospectus, and By-Laws, Santa Teresa de Jesus Silver Mining Company* (San Francisco: Turnbull and Smith, 1864), 13.

17. Hall, *Travels and Adventures*, 120–121, 232.

18. John R. Robinson Diary, 58, HM 62476, HL.

19. Hall, *Travels and Adventures*, 158, 140–42, 104; H. E. [Herman Ehrenberg], "The Old Babicanora Silver Mine of Sonora," *Weekly Arizonian* (Tubac), April 28, 1859.

20. Sylvester Mowry, *Arizona and Sonora: The Geography, History, and Resources of the Silver Region of North America*, 3rd ed. (New York: Harper and Bros., 1866), 45.

21. Hall, *Travels and Adventures*, 150.

22. Janin, "Report of Louis Janin, Jr.," 21.
23. Louis Janin Diary, 1863, HM 64294.
24. "New-Mexico," *New-York Daily Times*, May 11, 1852.
25. Robbins, *Colony and Empire*, 31.
26. "Gila Copper Mine," *Mining Magazine* 8, no. 5 (May 1857).
27. Stanley Buder, *Capitalizing on Change: A Social History of American Business* (Chapel Hill: University of North Carolina Press, 2009), 108–113.
28. [List of Mining Companies in Arizona, and their capitalization], *Mining Magazine*, 2nd series 1:1 (November 1859).
29. "Mining: It's Embarrassments and its Results," *Mining Magazine* 2, no. 6 (June 1854).
30. Diane M. T. North, *Samuel Peter Heintzelman and the Sonora Exploring and Mining Company* (Tucson: University of Arizona Press, 1980), 3.
31. On Heintzelman's background see North, *Samuel Peter Heintzelman*, 7–21; [Sonora Exploring and Mining Company], *Sonora—and the Value of Its Silver Mines; Report of the Sonora Exploring and Mining Co., Made to the Stockholders December 1856* (Cincinnati: Railroad Record Print, 1856), 7.
32. [Sonora Exploring and Mining Co.], *Sonora . . . 1856*, 5.
33. "Silver Mines in Sonora," *Mining Magazine* 3, nos. 5–6 (November–December 1854).
34. The choice of military men as managers often backfired. An agent for Samuel Colt, who took over the Sonora Exploring and Mining Co. in late 1859, observed of Heintzelman, "The habits of thought and manner characteristic of military men, render him rather unfit for the management of men on a frontier. I find that a spirit of dissatisfaction is generally prevailing among all classes at the mine. . . ." Robert Jarvis to Samuel Colt, December 6, 1858, C-47, box 19, Uncat. MSS.918, Jarvis-Robinson family papers (addition), Beinecke Library, Yale University [hereafter J-R Papers]; North, *Samuel Peter Heintzelman*, 41; [Sonora Exploring and Mining Company], *Third Annual Report of the Sonora Exploring and Mining Co., Made to the Stockholders, March 1859* (New York: W. Minns and Co., 1859), 3, 16; Robert Jarvis to Samuel Colt, September 8, 1858, [Santa Rita Silver Mining Company], box 19, J-R Papers; [Santa Rita Silver Mining Company], *Second Annual Report of the Santa Rita Silver Mining Company, Made to the Stockholders, March 19, 1860* (Cincinnati: Railroad Record Print, 1860), 4.
35. Samuel Colt, cited in Diane North, "'A Real Class of People' in Arizona: A Biographical Analysis of the Sonora Exploring and Mining Company, 1856–1863," *Arizona and the West* 26, no. 3 (1984), 261.
36. "Sonora Exploring and Mining Co., First Annual Report," *Mining Magazine* 8, no. 4 (April 1857).
37. Guido Küstel, *Roasting of Gold and Silver Ore, and the Extraction of Their Respective Metals, without Quicksilver* (San Francisco: Dewey, 1870); Küstel, *A Treatise on Concentration of All Kinds of Ores* (San Francisco: Mining and Scientific Press, 1868); Küstel, *Nevada and California Processes of Silver and Gold Extraction* (San Francisco: Frank D. Carlton, 1863).
38. "Silver and Copper Mining in Arizona," *Mining Magazine*, Second series 1, no. 1 (November 1859).
39. H[erman] Ehrenberg, "Valuable Statistics," *Weekly Arizonian* (Tubac), April 14,

1859; Ehrenberg, "A Valuable Table," *Weekly Arizonian* (Tubac), April 21, 1859; Ehrenberg, "Table of Distances," and "Table for Converting...," *Weekly Arizonian* (Tubac), April 28, 1859; Robert Jarvis to Samuel Colt, July 7, 1859, C-47, box 19, J-R Papers.

40. "Silver and Copper in Arizona," *Mining Magazine*, 2nd series 1, no. 1 (November 1859).

41. Guido Küstel, *Concentration of All Kinds of Ores*, 213.

42. John Shertzer Hittell, *Mining in the Pacific States of North America* (San Francisco: H. H. Bancroft and Co., 1861), 172–173.

43. "Silver and Copper in Arizona," *Mining Magazine*, 2nd series 1, no. 1 (November 1859).

44. H[erman] Ehrenberg to William P. Blake, ed., *Mining Magazine*, 2nd series (January 1860).

45. [Sonora Exploring and Mining Company], *Sonora...1856*, 43; [Sonora Exploring and Mining Company], *Third Annual Report...*, 20.

46. [Sonora Exploring and Mining Company], *Third Annual Report*, 7.

47. "Silver and Copper Mining in Arizona," *Mining Magazine*, 2nd series 1, no. 1 (November 1859).

48. *Weekly Arizonian* (Tubac), March 3–July 7, 1859.

49. Arizona Diary 1861, box 1, Raphael and Eliza Shepard Pumpelly Papers, Part I, The Huntington Library, San Marino, CA (hereafter Pumpelly Papers I).

50. Pumpelly, *Across America and Asia*, 15.

51. [Santa Rita Silver Mining Co.], *Second Annual Report...*, 17.

52. *Weekly Arizonian* (Tubac), July 21, 1859; [Santa Rita Silver Mining Co.], *Second Annual Report...*, 6.

53. [Santa Rita Silver Mining Co.], *Second Annual Report...*, 2.

54. Ibid., 12, 13.

55. Arizona Diary 1861, box 1, Pumpelly Papers I; [Raphael Pumpelly], "A Mining Adventure in New Mexico: A Real Experience," *Putnam's Magazine* 4, no. 22 (October 1869):494–502; Pumpelly, *Across America and Asia*.

56. Pumpelly, *Across America and Asia*, 15.

57. [Santa Rita Silver Mining Company], *Second Annual Report...*, 6. This portion of the report was signed by the president of the company, George Mendenhall, and was most likely from reports and letters written by Pumpelly, Charles Poston, and/or William Wrightson.

58. Pumpelly, *Across America and Asia*, 15–17, 19, 52–55; Arizona Diary 1861, box 1, Pumpelly Papers I; Receipt April 1861, folder Correspondence B-W, box 2, Raphael and Eliza Shepard Pumpelly Papers, Part II, The Huntington Library, San Marino, CA (hereafter Pumpelly Papers II).

59. Pumpelly, *Across America and Asia*, 26.

60. Hall, *Travels and Adventures*, 150; "Mines and Mining," *Weekly Arizonian* (Tubac), June 16, 1859.

61. Santa Rita Silver Mining Company, *Second Annual Report...*, 4.

CHAPTER TWO

INSTITUTING EXPERTISE

Mining Education in the United States

Herman Ehrenberg, mining engineer for the Sonora Exploring and Mining Company, observed in 1859, "The grand interest of the country being mining, this should be fostered by all means... [and] scientific and practical men, well acquainted with the manipulation and metallurgical treatment of ores, are indispensable."[1] This statement was not so much a compelling insight as it was a sober assessment of the realities of borderlands mining.

Ehrenberg, along with the handful of other mining engineers who worked through the borderlands region in the 1850s and 1860s, was a German-trained engineer. As such, he was a member of a distinct minority. The majority of the miners, prospectors, and assorted speculators who traveled to the region from the eastern United States at the time were, at best, self-taught mining men. University-trained experts such as Ehrenberg were few and far between in the United States in general and particularly uncommon in the borderlands. Although the southwestern United States experienced a couple of booms, such as that at Tombstone, notable for their pure, easily extracted ore, much of the mineral ore in the borderlands was of a lower grade that required significant processing before transport to market. Thus, mines that operated without good technical advice had little chance of success. Mine owners, investors, and operators were very aware of the need for technical workers, but in the 1850s and 1860s even members of Congress noted the shortage of such people. Senator William Stewart of the newly minted state of Nevada, for instance, noted in 1868 that "the number of truly scientific and practical men who have been engaged in the examination and working of our mines is extremely limited."[2]

Between the 1860s and the 1890s, however, the number of university-educated mining engineers increased significantly in the United States. In concert with other white-collar workers, most notably doctors, lawyers, and civil and mechanical engineers, mining engineers shared in the late-nineteenth-century drive for the standardization of professional expertise.[3] While fraternal and trade organizations played a role in the professionalization of mining engineering, they did not assume the gatekeeper role of groups such as the American Medical Association or the American Society of Civil Engineers. Rather, the development of a domestic system of dedicated mine-engineering education was of primary importance in changing the professional role of mining engineers between the 1860s and 1900. In turn, the professionalization and growth of mining engineering had a dramatic impact on the mining industry in the borderlands, providing regional mining engineers not only with a growing network of professional colleagues to call upon for assistance but also with the tools for asserting their authority that were either unavailable or impractical for engineers in Ehrenberg's generation.

To claim technical expertise, as mining engineers do, is to claim the authority to override social, political, or economic considerations in the service of a higher cause—that of a supposedly objective technical truth. Yet in the establishment of professional mine engineering, the relationship between the claim of technical expertise and the assertion of authority was not always so straightforward. Some of the earliest mining engineers to work in the Southwest and northern Mexico found that their academic credentials actually worked against their ability to assert authority. Later, some discovered that a mining engineer's need for academic and technical credentials, although seemingly paramount, was secondary to his need to demonstrate hands-on, local knowledge, as he established his plans as a technical advisor and on-site manager. This tension between professional credentialing and the necessity of demonstrating a strong practical knowledge of mining techniques and local circumstances became a driving force in structuring a U.S.-based system of mining engineering education between the 1860s and the early 1900s.

The Freiberg Network

Most mining engineers who worked in North America in the 1860s and 1870s were trained in Germany, for the most part at the Königliche Sächsische Bergakademie (Royal Saxon Academy of Mining) in Freiberg, Saxony, the foremost mining institution in Europe. Among the most important of the first generation of German-trained American engineers who spent a significant portion of their careers in the western United States and Mexico were Raphael Pumpelly, who worked for the Santa Rita Mining and Milling Company in

Tubac; James D. Hague, a major investor in southwestern and Mexican hard-rock mines; Guido Küstel, metallurgist and mining engineer for the Sonora Exploring and Mining Company, also in Tubac; all three Janin brothers, Louis, who spent the better part of 1863–64 examining mines in Sonora and Arizona, Henry, and Alexis; and Rossiter Raymond, longtime editor of the *Engineering and Mining Journal*.[4]

More often than not, graduates of Freiberg maintained lifelong connections to each other, corresponding about both personal and professional matters. Such friendships ensured that mining engineers in the new territories of the western United States remained part of a small network of educated and unusually cosmopolitan men. This network helped in obvious ways, enabling an easy flow of information about work opportunities and new mining methods, while taxing the Freiberg graduates with a certain social responsibility toward one another. In the early 1880s, for example, a Freiberg graduate named Price took a job in the West and was "broken down through softening of the brain or some other form of intellectual ruin." Two of Price's former classmates cared for him through his breakdown and took up a subscription among fellow alumni to send the unfortunate man back to his family in England. Louis Janin and James Hague, two successful Freiberg alumni from a previous generation, and neither of whom was a particular friend, together contributed more than a third of the expense of Price's trip home.[5]

At Freiberg, students split their time between classroom and "practical" studies in the field, a unique approach in mining education before 1864. Graduates of the Academy began their careers knowing the latest metallurgical techniques and theories of ore genesis. As with any credentialing process, the network of working professionals to which Freiberg provided access was as important as the education they received. Beyond the purely academic, a Freiberg education certainly bestowed several advantages on a mining engineer seeking a career in North America. Ambitious officials at marginally profitable companies hired trained mining engineers, hoping that their academic backgrounds would increase returns under exceedingly difficult conditions. As we have seen, this strategy was unsuccessful at borderlands mines in the 1850s and 1860s. Yet German-trained mining engineers continued to command high salaries in western mines. Their presence lent a veneer of respectability and authority to speculative ventures, which in turn conferred socio-technical authority as educated technical workers on the engineers.[6] Not only was the seemingly esoteric expertise of a Freiberg-trained mining engineer valued by investors, but among educated Americans German universities were considered the best in the world.[7] A young mining engineer who held a certificate or a degree from Freiberg was considered in certain circles superior to an engineer trained at other European institutions, such as the Royal School of Mines in London, or the École des Mines in Paris. A Freiberg graduate wanting a career in North America was in the happy position of possessing a level of education that could soothe the anxious investor who wanted to be sure he was hiring the "best" mining engineer available, while also having the kind of practical education that would be useful in wrangling actual mine workers.

Mining Engineers at the Frontier

Although "practically" trained mining engineers with German degrees such as Pumpelly and Küstel were hired on the basis of this nexus of their training and credentials, they were not always greeted with open arms at a mining site. Speculators, mine operators who fancied themselves (and sometimes were) very knowledgeable about mining processes and local conditions, and the apprentice-trained skilled workers who staffed mid-century mines were often very suspicious of the university-educated men brought in from outside to fix, upgrade, or reorganize a mine. Much of this suspicion stemmed from the inequities of class distinction: a Freiberg education was not inexpensive or easy to obtain, and German training signaled a life of privilege available to few Americans at the time. In the mid-nineteenth century, only the children of the elite could afford to spend three or four years living in Europe attending a mining school. Still fewer possessed the requisite language skills to study in Germany. Indeed, stories abound of Americans who cheated or got lucky on their language exams and then discovered they could not follow a technical lecture in German.[8]

Despite the challenges inherent to attending university in Germany, it was not uncommon, prior to the 1890s, for budding mine engineers to spend two or three years at an American university, such as Harvard or Yale, and then to "finish up" with coursework or a certificate from Freiberg or one of the other continental schools. Other German universities such as Göttingen, Causthal, and the Bergakademie in Berlin had respected mining schools. Further east, Schemnitz in Hungary had a similar course of study to Freiberg and attracted a number of American students. Some small number of students also did coursework in Paris at the École des Mines, but admission to French institutions required sponsorship of the French ambassador and was difficult for most foreigners to obtain. Almost no Americans studied at British institutions during this era. Indeed, the course of study at British mining schools in the mid-nineteenth century was considered by practitioners little better than the highly theoretical training a prospective engineer could get at Oxford or Cambridge, or, in the words of one graduate of the Royal School of Mines, "but a small degree better than none."[9]

Yet, although these early American mining engineers were almost exclusively the products of extremely privileged backgrounds, in another sense they were independent operators without the full support of a professional class.[10] They lacked institutional support within the mining industry, or automatic recognition of their education from colleagues. Each engineer had to negotiate his status on site by demonstrating either his worth or an appropriate humility before local knowledge.

The social privilege of mining engineers was thus a double-edged sword in the southwestern borderlands, where technical knowledge was measured not by degrees or scientific authority but by the ability of an engineer to build profits for his employer. As noted by many Freiberg alumni themselves, part of the problem with a mining education from Europe was the notably lower costs of extraction and transportation compared to the U.S. West. Many young engineers immediately fell afoul of their employers, overestimating the value of a mine by underestimating the expense of working in North America.[11] Although many people doubtless agreed with Senator William Stewart of Nevada that "an education at Freiberg is a guarantee not only to position and influence, but to the regard and confidence of the humblest miner," investors and managers also frequently complained through the 1860s that their university-trained mining engineers lacked technical skills.[12]

Captain J.W. Ruggles, for instance, mine manager at the Guazapares Mine in Chihuahua, fired his engineer due to the engineer's "dead, flat failure" to make the mine profitable. The engineer in question acknowledged that he "*might have* been roasting his ores too long; or that he might have been using too little salt." Neighboring mine operators and American agents in Chihuahua, meanwhile, agreed that the engineer "was not sufficiently acquainted with [the variety of ore at the mine] . . . by which means he ruined the whole concern," but they cast no aspersions on the engineer's general technical competence.[13] Ruggles, however, insisted he would have none of "*this* or that, or the other trifling excuse in extenuation of what appears to be [the engineer's] ignorance of the business."[14] Rather, it was in Ruggles's interest to place the blame for the failures at Guazapares on the shoulders of the mining engineer, rather than on external forces or on himself for hiring a man unacquainted with the local ores. In this instance, all observers, including the engineer himself, agreed that the engineer lacked practical knowledge that would help at the mine, but the vehemence of Ruggles's attack on his engineer underscores a problem that vexed mining engineers, particularly in the early days when there were very few in the region. Even if this engineer had roasted and treated the Guazapares ores properly, that alone was no guarantee that the company would turn a profit. Absent a balance sheet in the black, how could a mining engineer prove his value?

In practice, a German mining degree by itself did not immediately endear a mining engineer to his local associates in Arizona or Sonora, which compounded the engineers' difficulties in the field. John Hays Hammond, for instance, as a young mining graduate of both Sheffield Scientific School and Freiberg, was told outright by an acquaintance that he could not have a letter of recommendation since "engineers educated in the theory of mining had not been successful" in southwestern mining projects. Hammond, who later became one of the wealthiest and most influential mining engineers in the United States, learned from his mistake. He claimed to have obtained his first mining position as an assayer for George Hearst by forswearing the value of his engineering education altogether.[15] While Hammond played down his education, and Hearst may have proclaimed his mistrust of mining engineers, Hearst owed the long-term success of his mines on the Comstock and in Deadwood to the innovations of mining engineers, including those of Freiberg-educated Guido Küstel and Louis and Henry Janin.[16]

The value of a university degree in the eyes of working miners was also undermined by the strength of the apprenticeship system that thrived under the large population of Cornish immigrants who dominated the workforce in mid-century western and Mexican mines. These miners traditionally worked in teams that bid on labor contracts rather than negotiating a daily or monthly wage. Men advanced in pay grade and responsibility with seniority. Under the Cornish system, a manager was always a miner who had worked up through the ranks, not a college-trained engineer.[17] A college degree might have had some social value, but in a mining camp it was not worth as much as knowledge acquired in the field.

The Growth of an American Educational System

Given the shortage of substantive technical and managerial competence in the borderlands, plus the difficulty of acquiring a European education and the suspicion with which such education was greeted, a movement grew after the American Civil War to establish a national mining institution in the United States. Editorials in such disparate papers as the *Arizona Miner* and the *San Francisco Bulletin* argued forcefully that it was in the country's national interest to establish an American school of mines on the model of Freiberg or the Royal School of Mines in London.[18] In 1868, Senator Stewart of Nevada went so far as to introduce a bill in Congress to establish a National School of Mines to be located close to the major hard-rock mining regions of Colorado, Arizona, and Nevada. The arguments in favor of an American school included the nationalistic assertion, "We send our young men to the schools...in Europe, to learn that which they could much better learn at home, if we had institutions equally thorough and comprehensive." The observation that "the production of ore has decreased, as the amount of waste generated by mining has increased," and that the industry ought to focus on producing profits from low-grade ore, were also cited as evidence of a critical need for a national school of mines.[19] Two common objections raised to Stewart's proposal included a fear that establishing a Board of Mining to oversee technical standards would suffocate individual entrepreneurial investment, and the belief that taxpayers should not support the mining industry, since individuals, not the community, benefited from the industry.[20] The bill was withdrawn before the Senate could vote on it but not before Stewart gave an impassioned defense of the mining industry and mineral wealth in North America, insisting that American-trained Americans could do better mining work than men trained at "the great school at Freiberg."[21]

Lawrence and Sheffield

In the absence of an American school of mines, domestic options for training engineers were quite limited in the 1850s and 1860s, which is one reason Stewart's proposal generated so much attention in mining districts despite its lack of traction in the Senate. Prior to the Civil War, there were only a few institutions in the United States that trained engineers or provided a technical or scientific education. Rensselaer Polytechnic Institute, founded in 1823, and the Polytechnic Institute of Pennsylvania, founded in 1857, both trained civil engineers, but neither produced a significant number of graduates who worked in the mining industry.[22] Other schools that eventually had relatively large-scale mining programs, such as the California School of Mines at Berkeley and the Massachusetts Institute of Technology, had very few mining graduates before the twentieth century. The domestic schools that provided the greatest number of mining graduates during these early decades were the privately operated Lawrence Scientific School in Cambridge, Massachusetts, and the Sheffield Scientific School in New Haven, Connecticut. Of these schools, Sheffield eventually became the more prominent training ground for mining men, while Lawrence eliminated its mining program in 1879. But during the 1850s and 1860s the two were more alike than not.[23] Not only were the programs of study comparable but students and faculty alike bore the burden of affiliation with the least prestigious of the scientific fields: mining.

From the outset, both Sheffield, established in the early 1850s, and Lawrence, established half a decade earlier, granted degrees in civil and mechanical engineering. Dedicated mining programs were established a few years later, although it is noteworthy that in the mid-nineteenth century the mining program did not differ significantly from other courses in engineering. To modern eyes, all of these courses were astonishingly broad-based, with students taking more classes in fields such as botany, languages, rhetoric, and history than in mathematics, topography, and mapping.

While the mining students at these scientific academies were thought to be pursuing the least scholastic of the fields of study, all students at both Lawrence and Sheffield were considered the academic inferiors of their neighbors at Harvard and Yale. In later years, many students who enrolled in Lawrence after a couple of years at Harvard College came with weak academic records, and mining students were among the weakest of them all, suggesting that in the early years of dedicated mining education the brightest engineering students avoided mining because it was seen by their scientific peers as inferior.[24] In some ways, such evident disdain for the mining curriculum may have served graduates of Lawrence more ably in the field. If mine managers and workers greeted educated mining engineers with suspicion, perhaps claiming the credential of a geologist better helped establish one's technical authority. Regardless, the principle difference between the Lawrence curriculum and that at Freiberg remained the absence of a "practical" U.S. mining component.

In New Haven, faculty and students alike devoted more time to asserting the importance of dedicated training for engineers than did their peers at Lawrence. Eventually such effort resulted in Sheffield becoming a top engineering institution, while the prestige of Lawrence flagged among those in the mining business. But initially, demanding respect was an uphill battle for faculty and students alike. Entrance examinations were instituted as early as 1861, with mixed results. As one historian of Yale University explained, the entrance exams were useful as they:

> dissipated the danger that this school [Sheffield] would become a refuge for a certain class of students whose presence, while it added to the number of those receiving instruction, diminished the effective working of the instruction given.[25]

The trouble was that attempting to keep weak students out of Sheffield meant that the school lost revenue, and so the faculty also made provisions for admitting numerous "special" students who failed to pass the entrance exams. As a result, many students "who cared for only a nominal connection with the college in order to save themselves from being held responsible for their vigorous way of doing nothing at all" were admitted to the school in spite of the relatively difficult entrance exams. Sheffield courses thus had a very high attrition rate, and when students from Yale College, across the street, mocked "Sheff" students for being lackadaisical intellects who drank too much and were generally so dissolute they were "past praying for," students at Sheffield knew there was some truth to the characterization.[26] The combination of academic and social stigma attached to Sheffield undergraduates during the nineteenth century may have caused many of them to be reflexively defensive about their choice of (or need for) a career in mining. It almost certainly prepared them to forswear their academic credentials in the manner of young Sheffield alumnus John Hays Hammond.

The Morrill Land-Grant Act

Proponents of an American-based system of engineering education hoped that by broadening the educational options for prospective mining engineers beyond the rarefied bastions of Cambridge and New Haven, the passage of the Morrill Land-Grant Act in 1862 would revolutionize the domestic mining industry. The Act granted to each state a certain acreage of land to sell, determined from the 1860 census. Proceeds from the sale were to endow public "land-grant" universities devoted to teaching the principles of agriculture and the "mechanic arts." "Mechanic arts" was not clearly defined in the legislation, however, and most of the funds thus raised established institutions with relatively large agricultural programs and comparatively small faculties for engineering and the applied sciences. In a small number of cases, however, the Morrill Act did indeed advance the cause of engineering education.[27] Schools established by funds from the Morrill Act included the Massachusetts Institute of Technology and the Missouri School of Mines at Rolla, both of which eventually trained large numbers of mining engineers. The Morrill Act had some odd and unintended consequences, however. For instance, rather than establishing an agricultural college, Connecticut legislators diverted the semiannual interest payments from the Morrill Act into the operating fund of the otherwise privately funded Sheffield Scientific School. The mining curriculum there provided a basis for the argument that this was an appropriate use of funds.[28]

In the early years, the land-grant system was somewhat ad hoc. Many institutions only employed one or two engineering instructors, and the courses offered were contingent upon their knowledge. Yet, over time, the land-grant colleges developed what one historian of engineering education has called a "land-grant style." This "style" resonated with the particular job requirements for mining engineers very well, with an emphasis on the relationship between practical and theoretical knowledge and a significant proportion of class time devoted to field or laboratory work.[29] Although the Morrill Act did not create a network of mining colleges, as proponents of a national mining academy hoped, it did support a style of education that was closer to the Freiberg model than to other European institutions or Lawrence or Sheffield in the 1850s and 1860s. The land-grant colleges, by providing a public source of funding for training engineers, contributed to the overwhelming emphasis on *practice* that came to dominate American engineering education by the end of the nineteenth century.

Of more immediate importance to the mining industry, however, the Morrill Act dramatically and immediately increased the number of American institutions providing engineering education. In 1860, there were seven engineering colleges in the United States. By 1872, there were seventy.[30] Fewer than 300 men received engineering degrees in the U.S. in the two decades preceding the Civil War. Over the next ten years, that number quadrupled. Ten percent of those new engineering graduates—approximately 120 men over the course of a decade—were mining engineers with specialized degrees.[31] Despite the relatively small number of men who possessed mining degrees in the 1870s and early 1880s, as engineering education became progressively more available, investors and financiers began to expect that anyone who called himself a mining engineer would have a college degree. The numbers did increase, with one analyst counting nearly 900 mining degrees awarded between 1867 and 1892, primarily from Columbia, but with MIT a distant second.[32] In the long run, these American engineering colleges changed the demographics of the profession of mine engineering. By the turn of the century, more than 10,000 students were enrolled in engineering colleges. And while the private institutions remained the most prestigious schools for mining, the preponderance of land-grant colleges enabled many more men to take at least *some* college courses. Indeed, the majority of mining graduates in 1900 came from the land-grant system, either the established mining schools at Rolla, Missouri, and Houghton, Michigan, or from institutions established in the territories (and therefore not subject to the Morrill Act) such as New Mexico Mines (Socorro), established in 1889, and Arizona School of Mines (Tucson). The latter began instruction as part of the territorial university in 1891 but did not grant degrees exclusively in mining until 1915.[33]

The Columbia School of Mines

Over time, the Morrill Act revolutionized the engineering profession within the United States, but in the short term, the most significant challenge to the dominance of Freiberg alumni in the nineteenth-century mining industry was the opening of another private mining institute in 1864—the Columbia School of Mines in New York City. While all mine engineering curricula were based on a broad offering from the liberal arts in addition to more obvious mining and engineering classes in geology, chemistry, trigonometry, calculus, physics, and German, what made Columbia stand out was its Summer School of Practical Mining.[34] This program differed from the practically oriented land-grant system of education in that the practicum was the central feature, around which all other aspects of the Columbia mining curriculum were based. Consciously modeled on the practical curriculum at Freiberg, the School of Mines required all third-year mining students to spend the summer as apprentices in a working mine, mostly in the copper mines near Lake Superior in Michigan's Upper Peninsula.[35] Neither Lawrence nor Sheffield offered anything comparable. Some of Columbia's early success undoubtedly stemmed from the belief, common in the North American mining community, that "our schools are not practical enough, [and] that practical miners do not find the graduates capable of doing what they profess."[36]

Columbia had strict entrance requirements and also provided relatively good opportunities for scholarship students, making it a top choice for upwardly mobile prospective engineers. Its relative success in usurping the dominant position of the European academies, principally Freiberg, as the premier institution for budding mining engineers also owes a debt more generally to the direction of American scientific study in the late nineteenth century. Historians of science have puzzled over the predominance in America of experimental and observational scientists over theoretical

Students at Columbia University School of Mines, 1888. *The Miner* yearbook, New York: Columbia School of Mines, 1888, frontis.

and mathematical thinkers prior to 1900, a situation that seems to point to a certain backwardness.[37] Mining, a field of study that privileges fieldwork and the application of theoretical knowledge for immediate and concrete financial gain, fit the American model. The Columbia summer school was designed explicitly to bring the social and the technical aspects of mine engineering together and confer the practical experience that would legitimize graduates in the eyes of grizzled old miners. In practice, it served to underscore the class distinction between engineers and working miners and gave the engineering students the tools they would need to negotiate their status at a mining camp.

Growing up in Toledo, Ohio, William Field Staunton was so impressed by his neighbor's son, a mining engineer named John A. Church, who worked at the time in Tombstone, Arizona, that he decided to follow in Church's footsteps to Columbia.[38] Staunton, who eventually became the president of the Congress Mine, one of the largest gold mines in Arizona, recalled to the end of his life the social stigma he felt as a student in New York. Although the son of a successful railroad engineer, as a student he judged himself cash-strapped, unsophisticated, and Midwestern compared to his classmates. Staunton's term of enrollment in the Summer School of Practical

Mining, however, demonstrates the effect that a Columbia education had on the relationship between even a middle-class Midwestern boy and the "practical" miners whom he would direct throughout his career.

In the summer of 1881, Staunton's Columbia class worked at a copper and iron mine in the Keweenaw Peninsula of northern Michigan. When he first arrived, Staunton made an error typical of first-time miners and forgot to extinguish the candle in his hat before it guttered. As the candle flickered out, it melted the resin that attached it to the front of his cap, and the wax, resin, and end of the wick all stuck to Staunton's forehead. His friends had to shave his hair off to remove the mess, leaving a readily recognizable symbol of inexperience for all to see.[39] A rube in New York, Staunton also found himself to be a tenderfoot underground, although he was able to parlay that early naïveté into a story that showed him capable of modesty and able to play by the rules of the underground workers.

Staunton's troubles at the summer school did not end with a shaved head. Later in the summer, and more comfortable with life underground, he heard that some of the miners preferred to leave the mine by climbing directly up an 800-foot ladder rather than by walking out through the tunnel. Staunton reasoned that this shorter climb would be a faster, better way to exit and followed a group of men up the ladder toward the surface. Unaccustomed to such physical labor, Staunton soon exhausted himself and, losing sight of the miners ahead of him, clambered hand over hand in the pitch dark after his candle burned out. He recalled,

> At last…I saw a faint glimmer, like a fire-fly, seemingly miles below me in the shaft. The light grew, and then I could hear the steps of men on the ladders coming up. On few occasions in my life have I felt a more joyful sense of relief. They found me, perched like a squirrel, at the top of a ladder which had projected farther past one of the landings than the others, and which I had passed in the dark. The miners seemed to find the situation hugely amusing, a feeling I did not share at the moment.[40]

The absurdity of Staunton's situation is apparent. The underground workers clearly understood that the "college boys" would one day be the supervisors and managers of works just like their own. What a pleasure to find a young engineer-in-training "perched like a squirrel" on a ladder, unable to keep his candle lit or to find the mine exit without recourse to their superior local knowledge.

Staunton also had something to gain from telling this story. His rookie blunders were shorthand for explaining that he had already made mistakes in underground mining and was unlikely to make such errors again. That the stories have a comical side also suggests that the youthful Staunton may have been overly exuberant, but he was never, even as a young man, prone to major errors of judgment. He therefore approached practical mining as an experienced hand rather than a green college boy. *Engineering and Mining Journal* once explained, "The principal advantage gained by this manual work [in the Columbia summer school] was the braking [*sic*] of the ice between the student and the miner, placing them in pleasant relations and on common ground, the student for the time becoming, to all intents and purposes, the miner's 'butty,' and to be treated and instructed as such."[41]

Although it seems unlikely that the class divide between the principally foreign-born, apprentice-trained miners and the predominantly Eastern, university-educated mining engineers could really be bridged, Staunton's stories indicate that he really *was* treated as a miner's "butty," and that he called on this experience to legitimize his authority for the rest of his career. Columbia graduates thus began their careers with a real advantage over their other colleagues: while a graduate of Sheffield, for instance, would have to find a way to work underground for a couple of years to gain Staunton's experience—a project few, if any, Sheff men undertook on their own—a Columbia graduate was understood by his peers and potential employers to have a worker's insight into underground work. Columbia graduates themselves benefited at the beginning of their professional careers by learning how tenuous their claims to authority would be underground. Although even a young engineer might have plenty of theoretical knowledge of how and what to mine, he would still have to negotiate with the miners to make the work happen.

Yet even Columbia's famous practicum was not actually the same as an apprenticeship. Many people in the industry—engineers and owners alike—felt there was simply no substitute for extensive underground experience. This inexperience was industry shorthand for criticizing engineering work as sloppy or poorly conceived. An anonymous engineer's report on the Santa Elena mining property in Sonora, for instance, lamented the work done by Eben Olcott, who was employed to manage the property in 1881, shortly after he graduated from Columbia. "Olcott carried out [his work] in a most elaborate and costly manner," the report stated. "Properly speaking, [he produced] no cut, it was simply an enormous out-crop of quartz that was carried out." In other words, Olcott failed to engineer the mine. Rather, he grabbed a chunk of surface rock and called it mining. Even more egregious, by the account of the report's author, Olcott chose the wrong process for milling ore extracted by a previous manager and allegedly lost 60 percent of the ore's value.[42] This failure, the report's author implied, was a direct result of Olcott's lack of practical knowledge and his reliance on book-learning over common sense. If Olcott did indeed commit these errors (and there is no way to verify the claim either way), they were costly mistakes, both in terms of lost revenue and, presumably, in labor expended to remedy the mistakes.

In fact, Eben Olcott went on to a remarkably successful career as a partner in a prosperous consulting firm based in New York and a well-known expert on southwestern, Mexican, and Latin American mining projects, serving at one point as the president of the American Institute of Mining Engineers. This professional success stemmed in no small measure from the combination of his well-respected Columbia degree and to the many mining projects in Mexico he managed as a young man, including the Santa Elena.[43] Yet as a young engineer, the Columbia degree could not fully overcome the suspicion on the part of other engineers that any questionable decision he made stemmed from overreliance on theoretical knowledge. Olcott's situation points to a curious aspect of the Columbia model of field education. Although it was instituted to further the *mining* knowledge of students, the summer school was actually most effective in alerting young mining engineers to the fact that they would have to negotiate their status at the mine and that it would not be self-evident to the workers that a mining degree made a man an engineer.

Overall, however, the success of the Columbia system was apparent to people in the industry, and by 1880 other institutions followed suit. The Massachusetts Institute of Technology, for instance, installed a "mining lab" on its campus in Cambridge, and the California School of Mines instituted a requirement that mining students use their summer vacations to visit area mines and metallurgical works. Until the late 1890s, however, more engineering graduates of Columbia worked in the mining industry and in the related field of metallurgy than did graduates of any other American university.[44]

Western Mining Institutions

By 1900, when engineering education had become established in the United States, institutions in the newer states began to outstrip Columbia's influence. The California School of Mines was the biggest. In 1900, it claimed an enrollment of over 300 students in various aspects of mining, many of whom were hoping for work in the newly lucrative markets of Australia and South Africa, but a portion of whom stayed closer to home. The turn of the twentieth century was the high point for California enrollment. By 1915, it had dropped off significantly. The Colorado School of Mines also took off during the early years of the twentieth century and continued to grow in scale of program and scope of studies, training scores of engineers for work with the refractory and difficult minerals of western North America. A major benefit of these western schools versus the older, and sometimes more prestigious, institutions of the East was that a formal, practical course such as the Columbia summer school was unnecessary. Come summer, students were simply encouraged to go forth and find low-wage work at a mine. One California professor, indeed, regularly arranged for his students to spend summers pushing wheelbarrows at local mines for $1.50 a day, practically starvation wages even at the turn of the century. New Mexico went one step further, and simply purchased a mine next door to the institute in Socorro for the practical edification of its students.[45]

The Effect of Educational Changes on the Profession

One of the most noteworthy features of the history of engineering education in the U.S. in general, and of mine engineering education in particular, is its evolution from a broad-based classical education with a heavy emphasis on languages and mathematics to a specialized course of study in physics and chemistry. In the early 1870s, every student at Sheffield, for instance, took a three-year course in either mechanical or civil engineering. Students who wished for a degree in mining took a fourth year of metallurgy. Regardless of major, each student was required to study two full years of both German and French, as well as three years of drawing, geometry, physics, differential and integral calculus, some courses in botany, English composition, and geography, as well as specialized courses in mechanics and stone cutting (for civil engineers) and steam engines (for mechanical engineers). In addition, all were required to study several semesters of elocution, although some faculty believed this should be an optional requirement.[46] Sheffield actually eliminated many of the language requirements in the 1880s, a move that in retrospect made the curriculum more modern than that of other private scientific institutions.

By the 1890s, the Sheffield faculty was again concerned that the entrance requirements were too low and began to demand that students read either French or German upon entrance and have enough command of Latin to read Caesar in the original.[47] By 1901, the stricter entrance requirements remained in place, but almost all of the "frills" in earlier Sheffield course requirements had been eliminated for mining engineers, including botany and elocution. These were replaced by the more professionally relevant studies of thermodynamics, mineralogy, crystallography, hydraulics, chemistry, and machine design. While calculus remained in the curriculum, geometry, algebra, and trigonometry were relegated to the entrance requirements, and advanced study of these fields was no longer mandated. Such changes were intended "to enable the men to have a course leading to more practical results in their professional work." Notably, by 1906, Sheffield had instituted an optional course in Spanish, a far more useful language than French or German for the many Sheffield graduates who worked at mine sites in the southwestern United States and throughout Latin America.[48]

To some degree, this shift in educational focus reflects changes in mining methods. The non-selective methods of the twentieth century were heavily dependent on an understanding of the chemical attributes of an ore. The physical manifestation of an ore-body was of secondary importance, as the quantity of rock removed from the ground was so great. The great changes in the coursework of prospective mining engineers, however, also serve to underscore the ways in which the working role of mining engineers changed through the first decades of the profession. In the 1850s and 1860s, mining engineers were classically trained advisors, whose knowledge of geology and chemistry was far superior to that of most practically trained miners. A half century later, the curriculum had changed such that engineers could more readily serve as critical technical experts, advisors, and managers.

By 1900, mine engineering professionals were wholly invested in the American university system, while remaining committed, in principle if not in practice, to the idea that a "practical," or apprenticeship-centered, education was in some sense superior to a "theoretical," or university-centered, education. As engineer George Packard expressed in a letter to the editor of *Mining and Scientific Press* in 1907, "There has never been such a demand for the college graduate as there is today." Indeed, Packard continued, it "unfortunately is not necessary for the present graduate 'to start at the bottom.' It would be much better for him if it were."[49]

In 1907, Packard was nearing the end of his career, but the ambivalence he expressed concerning the replacement of practical mine work with college credit was not simply generational. Rather, mining engineers as a group persistently expressed discomfort with the perception that they were among the elite and noted with concern that the best place to learn mine engineering was not in the classroom but underground.[50] As late as the 1930s, the alumni bulletin for the Colorado School of Mines ran an article discussing the concern of School of Mines graduates that they would be accused of "looking for a 'white-collar job'" because they were "afraid to soil [their] hands in manual labor."[51] The publication naturally concluded, "No one can accuse the young alumni of the Colorado School of Mines of such an attitude." Yet the fact that this fear could be stated so baldly is an indication of how troubled the profession continued to be with the idea that a university-trained engineer was less "practical" than a working stiff down in the mine.

The introduction of a widespread technical education for mining engineers did more than create a more uniform class of professionals; it also affected an engineer's advancement throughout his career. As it became easier to get a technical education, career prospects dwindled for men without university certification. Halfway through 1913, the *Bulletin of the American Institute of Mining Engineers* began to publish the work histories of men nominated for membership. While these listings include only a small sample of mining professionals, the listed individuals were among the most professionally active, and can be taken as indicative of the demographics of the profession as a whole. The vast majority of the nominees attended engineering institutes and held graduate (E.M.) and/or undergraduate degrees in mining. Only a handful of the nominees had not attended university. Of these, most were between the ages of thirty-five and fifty-five, with two or three decades of experience as superintendents or overseers before being nominated for AIME membership.[52] During the 1870s and 1880s, a man known to have good "practical" skills could readily find employment as an expert, and the ages of the men nominated in 1913 indicates that they probably began working in mining in the late 1880s or early 1890s. The nominees with college degrees were significantly younger, most under thirty with no more than eight years' work experience. More often, these young men were working at only their second or third professional position and had been out of college for fewer than four years.

By the twentieth century, possession of a mining degree was presumed to be the decisive factor in hiring decisions, even when hiring managers declared education (or the lack thereof) to be irrelevant. In 1902, for instance, Courtenay DeKalb, a Columbia-trained mining engineer in charge of operations at the Fernando Mining Company, was pressured by the company owner, Colonel Livermore, to hire his son, Tom. When DeKalb expressed reluctance to take Tom on as assistant engineer, the Colonel assumed that DeKalb's reluctance stemmed from the fact that young Livermore "ha[d] not received a technical training in the schools." DeKalb objected to this characterization, protesting that he did not care whether Tom had a college degree or not. The Colonel continued to demand that DeKalb nonetheless hire his son, and DeKalb continued to refuse, asserting that Tom was unreliable and knew nothing about geology. Had Tom Livermore completed coursework in mining, he would have studied geology, thus mitigating at least one of DeKalb's reservations. Had Tom completed a college degree, he might also have demonstrated some reliability, or at the very least the ability to complete a task once he began it. Despite DeKalb's protestations that he did really not care whether Tom Livermore had a college degree or not, Colonel Livermore was clearly correct to think that his son's lack of a degree materially damaged his professional chances.[53]

Over less than a generation, in concert with the opening up of the educational landscape for mining engineers and the increasing specialization of mine engineering coursework, the employment field for mining engineers changed dramatically. By the 1890s, engineers who themselves graduated from college in the late 1870s or the early 1880s—barely a decade after the Columbia School of Mines was established—were already making excuses to their superiors on behalf of informally trained coworkers whom they wanted to promote. In 1908, when William Field Staunton resigned from his position as superintendent of a group of mines in Tombstone, he recommended as his successor a man without a university background, Bert Macia, whom he described as "a very good practical man."[54] Managing to sound both defensive and paternalistic, Staunton explained to his superiors, "[Macia] has the advantage of being a practical miner and the promotion would undoubtedly stimulate him to do his best." Furthermore, Staunton asserted, "I believe in promotion from the ranks where the man has good stuff in him even though some of the embellishments may be absent."[55]

What Staunton coyly referred to as "embellishments," were nothing of the sort. Rather, he was indicating to his superiors his understanding that a person who did the work of a mining engineer—which, when Staunton resigned from Tombstone, included supervisory positions such as his—ought to possess a degree from an engineering institution. By the twentieth century, institutional affiliation could stand in for practical, "ground-up" knowledge, even in borderlands mines in some of the least institutionalized and most antiestablishment mining districts in North America a mere four decades earlier.

In the 1850s and 1860s, hard-rock mining in the American West and northern Mexico was primarily the work of independent prospectors, apprentice-trained miners, and autodidactic speculators. University-trained engineers came up through a system dominated by the German mining academy at Freiberg, although they were, for the most part, Anglo-Americans from the northeastern United States. After the American Civil War, the Columbia School of Mines led the way in establishing a new mining curriculum in the United States that was a hybrid of theoretical and practical training, modeled on the system at Freiberg.

The success of the Columbia model, in combination with the establishment of several new dedicated engineering institutions in the 1860s and 1870s, caused the general faculty at engineering colleges to take mining seriously as a field of study separate from civil or mechanical engineering. Narrowing curricular focus in mining programs through the late nineteenth century coincided with a tremendous increase in the number of students educated at the newly established engineering and technical colleges. At the same time, the mining industry came to need workers who could extract ever-lower grades of ore, and who therefore had advanced training in metallurgy and chemistry. By the turn of the twentieth century, university-educated mining engineers were critical workers in the mining industry. No longer outside consultants, mining engineers were now an integral part of any mining venture.

Notes

1. [Herman Ehrenberg], "The Apache Treaty," Tubac (AZ) *Weekly Arizonian*, June 2, 1859.

2. [William Stewart], "Speech of the Hon. William M. Stewart, of Nevada, on the Bill to Establish a National School of Mines, Delivered in the Senate of the United States, January 16, 1868" (Washington, D.C.: F. and J. Rives and Geo. A. Bailey, 1868), 11.

3. On the rise of a professional class in the United States, see Chandler, *The Visible Hand*; Noble, *America by Design*; and Wiebe, *The Search for Order*, especially ch. 5.

4. Spence, *Mining Engineers*, 25; and J. Ross Browne, *Adventures in the Apache Country: A Tour through Arizona and Sonora, with Notes on the Silver Regions of Nevada* (New York: Harper and Brothers, 1871; reprint, New York: Promontory Press, 1974), 217.

5. James D. Hague to Mr. Booream, February 20, 1883, and James D. Hague to Henry Janin, October 29, 1883, folder L-10, JDH. Wallace Stegner fictionalized this episode in *Angle of Repose* (Garden City, NY: Doubleday, 1971).

6. [Sonora Exploring and Mining Co.], "Fourth Annual Report," 5, 14; [Santa Rita Silver Mining Co.], "Second Annual Report" (Cincinnati: n.p., 1860), 10; and J. W. Ruggles to Board of Trustees, Guazapares Mining Co., April 14, 1865, BC 2071, Henry D. Bacon collection (hereafter HDB), The Huntington Library, San Marino, CA.

7. Roger Geiger, *To Advance Knowledge: The Growth of American Research Universities* (New York: Oxford University Press, 1986); Peter Meiksin and Chris Smith, eds., *Engineering Labor: Technical Workers in Comparative Perspective* (London: Verso, 1996), 4.

8. Spence, *Mining Engineers*, 27; William Randolph Balch, *Mines, Miners, and Mining Interests of the United States in 1882* (Philadelphia: Mining Industrial Publishing Bureau, 1882), 345.

9. Spence, *Mining Engineers*, 30–33.

10. Spence, *Mining Engineers*, 25; John Hays Hammond, *The Autobiography of John Hays Hammond*, vol. 1 (New York: Farrar and Rinehart, 1935), 66.

11. Spence, *Mining Engineers*, 35.

12. [Stewart], "Speech," 11. For typical complaints about untalented engineers and managers, see folders BC 2071, BC 2035, BC 1974, HDB; and Robert Jarvis to Samuel Colt, May 26, 1859, folder C-47, box 19, J-R Papers.

13. J. W. Ruggles to Pres, Sec'y Board of Trustees Guazapares Mining Co., April 14, 1865, BC 2071; William Jennings to John Heard, May 23, 1865, BC 2036, both HDB.

14. J. W. Ruggles to Board of Trustees, Guazapares Mining Co., April 14, 1865, BC 2071, HDB.

15. Hammond, *Autobiography*, 83–85.

16. A. D. Hodges, "Amalgamation at the Comstock Lode, Nevada," *Transactions AIME* 19 (1890); *Mining and Scientific Press* (May 21, 1910).

17. Spence, *Mining Engineers*, ch. 3; Lankton, *Cradle to Grave*, 58, 60–61, 65.

18. Prescott (AZ) *Arizona Miner*, "The Proposed Mining College," March 14, 1866; "Opinions of the Press and of Eminent Public Men on the Importance of Our Mineral Resources and the Advantages to be Derived from the Establishment of a National School of Mines" (Washington, D.C.: Wm. H. Moore, 1868), 12.

19. "Considerations in Reference to the Establishment of a National School of Mines as a Means of Increasing the Product of Gold and Silver Bullion" (Washington, D.C.: Intelligencer Printing House, 1867), 5; "Opinions of the Press...", 4; [Stewart], "Speech."

20. Prescott (AZ) *Arizona Miner*, "The Proposed Mining College," March 24, 1866; (Stewart), "Speech"; "Opinions of the Press. . . ."

21. Wm. Stewart, *Congressional Globe*, 556–561; January 16, 1868, 40th Congress.

22. Noble, *America by Design*, 22; Spence, *Mining Engineers*, 37.

23. In the early twentieth century, Lawrence and Sheffield were both absorbed by their neighboring institutions, respectively, Harvard and Yale Universities. Samuel Eliot Morison, *The Tercentennial History of Harvard College and University, 1636–1936*, vol. 1, *The Development of Harvard University since the Inauguration of President Eliot 1869–1929* (Cambridge, MA: Harvard University Press, 1930), 415.

24. Records of Students in the Lawrence Scientific School, 1888–1911, Harvard University Archives, Cambridge, MA.

25. Thomas Raynesford Lounsbury, *Sheffield Scientific School, 1847–1879. An Historical Sketch*... (New Haven, CT: Press of Tuttle, Morehouse and Taylor, n.d.), 7–9.

26. Loomis Havermayer, *Sheff Days and Ways: Undergraduate Activities in the Sheffield Scientific School, Yale University, 1847–1945* (New Haven: 1958), 8–11.

27. James Gregory McGivern, *First Hundred Years of Engineering Education in the United States (1807–1907)* (Spokane, WA: Gonzaga University Press, 1960), 93.

28. Lounsbury, *Sheffield Scientific*, 7–9, 20. The Connecticut legislature was inspired by the example of New York State, which used its much more substantial land-grant money as a matching grant to persuade Ezra Cornell to establish Cornell University.

29. Bruce Seely, "Reinventing the Wheel: The Continuous Development of Engi-

neering Education in the Twentieth Century," in Alan Marcus, ed., *Engineering in a Land-Grant Context: The Past, Present, and Future of an Idea* (West Lafayette, IN: Purdue University Press, 2005), 164–165.

30. McGivern, *First Hundred Years*, 88.

31. McGivern, *First Hundred Years*, 73–74, 90. The numbers of pre-Civil War engineers are based on graduation statistics from the earliest engineering colleges in the United States: Rensselaer Polytechnic Institute, University of Michigan, Dartmouth College, Sheffield Scientific School, Lawrence Scientific School, and Union College. No women received degrees in mine engineering in the 1860s.

32. Samuel B. Christy, cited in Spence, *Mining Engineers*, 40. Christy does not include either Sheffield or Lawrence in these numbers. Lawrence would likely be low on the list, as its mining program was disbanded in 1879, but Sheffield was still educating miners at a decent rate into the twentieth century. Spence himself rightly underscores the importance of Columbia for western mining but ignores the significance of Sheffield in the nineteenth century.

33. Layton, *Revolt of the Engineers*, 4; Bruce Seely, "Reinventing the Wheel,"163; Thomas Thornton Read, *The Development of Mineral Industry Education in the United States* (New York: AIME, 1941), 109; 112-113.

34. Balch, *Mines*, 271–343.

35. Ibid., 297.

36. Ibid., 344.

37. John W. Servos, "Mathematics and the Physical Sciences in America, 1880–1930," *Isis* 77, no. 4 (1986):611–613.

38. William Field Staunton, "Memoirs of William Field Staunton: The First Fifty Years, 1860–1910," p. 21, box 1, William Field Staunton papers, AZ 152 (hereafter WFS), University of Arizona Libraries, Special Collections, Tucson, AZ.

39. Staunton, "Memoirs," 25, box 1, WFS.

40. Ibid., 27, box 1, WFS.

41. *Engineering and Mining Journal* 30, no. 9 (11 August 1877), cited in Balch, *Mines*, 297.

42. "The Santa Elena Gold Mine," November 5, 1894, folder P-117, JDH.

43. Eben E. Olcott to Arthur Macy, December 2, 1885; Olcott to John Brooks, November 19, 1885; "Report on Certain Claims near Alamos, Sonora," January 2, 1907. Olcott, Corning, and Peele, no. 3, all BV Olcott, New York Historical Society, New York, NY (hereafter BV Olcott).

44. Balch, *Mines*, 283, 307, 310, 318, 342–43; Spence, *Mining Engineers*, 40. A survey of graduates of the Columbia School of Mines, conducted over the winter of 1881–1882, found that nearly seventy-seven percent of Columbia's graduates were employed in the mining industry in some capacity: as engineers, metallurgists, or assayers. *School of Mines Quarterly* 3, no. 3 (March 1882):242.

45. Spence, *Mining Engineers*, 44–46; Read, *Development of Mineral Education*, 82–91.

46. "Programme of the Sheffield Scientific School of Yale College for the College Year 1873–1874" (New Haven: Tuttle and Morehouse, 1873); [Governing Board Minutes] January 12, 1874, folder 55, box 6, RU 819.

47. "Programme of the Sheffield Scientific School of Yale College for the College

Year 1880–81" (New Haven: Tuttle, Morehouse and Taylor, 1881); Governing Board Minutes, May 9, 1892 and May 30, 1892, folder 55, box 6, RU 819.

48. Governing Board Minutes, March 11, 1901; May 22, 1905; and February 5, 1906, all folder 56, box 6, RU 819.

49. George Packard, "Mining Schools and Their Graduates," *Mining and Scientific Press* 95 (August 10, 1907):173–174.

50. *Engineering and Mining Journal* 36, no. 10 (September 8, 1883):142.

51. "Not Afraid of Work," *Colorado School of Mines Magazine* 21, no. 11 (1931):8.

52. "Candidates for Membership," *Bulletin of the American Institute of Mining Engineers*, nos. 78–81 (June–September 1913).

53. Colonel Livermore to Courtenay DeKalb, September 16, 1902; Courtenay DeKalb to Tom Livermore, October 19, 1902; Tom Livermore to Courtenay DeKalb, November 10, 1902; all folder 92, box 9, Courtenay DeKalb Papers, MS 1176, Arizona Historical Society, Tucson, AZ.

54. Staunton, "Memoir," p. 226, box 1, WFS.

55. William F. Staunton to Henry Robinson, December 26, 1908, folder 1, box 4, WFS.

CHAPTER THREE

WESTERING EASTERNERS

Class, Masculinity, and Labor

In the 1870s and 1880s, Mary Hallock Foote, the wife of Columbia-educated mining engineer Arthur De Wint Foote, traveled with her itinerant husband to several major mining sites, including Leadville, Colorado; Grass Valley, California; and Michoacán, Mexico. Although socially and professionally well connected, Arthur Foote did not enjoy particular success in his career, and for much of the late nineteenth century his wife supported the family as an author and illustrator for literary magazines such as *Putnam's*, *Atlantic Monthly*, and *Century Illustrated Monthly Magazine*. The success Mary Hallock Foote found in the literary world was due to her ability to interpret for a predominantly northeastern, educated audience the exotic nature of the places her husband worked. In a posthumously published memoir, she described her husband's circle of friends as "professional exiles," "remarkable men, cultivated, traveled, [and] original."[1]

While the depiction of her husband as an "exile" probably says more about Hallock Foote's own mixed feelings about living apart from her literary friends in New York than her husband's attitude toward mining camps and miners, the gulf to which she alludes between the class background of university-educated mining engineers and almost everyone else in a mining camp was real.[2] Class affiliation underpinned the professional identity of mining engineers through the late nineteenth century, not only in mining camps but also in the popular perception of mining engineers and in their self-presentation in the drawing rooms and boardrooms of New York and San Francisco.

Through the early years of the twentieth century, as mining engineers were increasingly drawn from the ranks of college-educated men, their class status was clear: they were the sons, brothers, and classmates of mining investors and capitalists. Historians have long noted the fundamental conflict between engineering professionalism and the business interests that engineers served. Could mining engineers ever truly separate expertise from collusion with the company line? Or were they businessmen first and principled advisors second? In the mining industry, these roles were more entangled than they were for civil or mechanical engineers, who rarely stood to make money apart from their salaries. But the line between profit and expertise was blurred in mining: a weak engineer could be in the right place at the right time and strike pay dirt. In the case of a successful mine—or a mine that was *reputedly* successful—an engineer could pocket a percentage of profits and additional consulting fees along with his initial salary and expenses. Conversely, a talented and honest engineer might repeatedly disappoint his employers by offering frank, and discouraging, assessments of mine sites. Because of such haziness, other engineers considered those in mining to be the most ethically compromised and least "professional" engineers.[3] Sometimes the only evidence that a mining engineer did a good job was when a mine turned a profit. Even if they served only as employees and were not actually stockholders or in upper management, consulting engineers were deeply interested in the fortunes of the companies for which they worked. Of course, many mining engineers were stockholders *and* employees of the Board of Directors. Such overlapping identities were a fraught ethical issue for the profession into the early decades of the twentieth century.

Yet, since success for a company meant success for an engineer, "good business" and "principled technical professional" were not really separate things. Rather, technical knowledge in service to profit was the raison d'être of the profession, and mining engineers worked hard, as individuals and as a group, to present themselves as competent technical professionals rather than as spokesmen for a company or the industry as a whole.[4] For mining engineers working in the borderlands, the distinction between company man and technical man was particularly complicated, as the "company" was not readily present to protect them and enforce their authority. Generally, central offices and investors were far off site, usually in centers of capital such as New York or San Francisco. When on site at a mine, even a very junior engineer was usually among the highest-ranking company officials present.

Consciousness of their status pervaded the workplace interactions of mining engineers and, in concert with their status as technical workers, underpinned their professional identity. Mining engineers negotiated their bureaucratic status most notably by embracing the westering frontiersman as a self-image. This quintessentially American identity both reified their racial status as white men and had a fundamental classlessness that could paper over the obvious distinction between engineers and the mass of mine workers.

Mining engineer John Greenway exemplifies this tradition. A former Rough Rider and personal friend of Theodore Roosevelt, Greenway presented himself as a straight-talking frontiersman and citizen of Arizona while amassing significant personal wealth and influence. He was so patriotic that he condemned fellow Americans for investing in mines south of the border, simultaneously rounding up his fellow engineers-cum-capitalists, L. D. Ricketts and James Douglas, to join with Theodore Roosevelt and himself adventuring on the frontier and among the "cannibals" near Guaymas.[5] As the superintendent of the Calumet and Arizona Mining Company in the early twentieth century, Greenway was one of the most powerful men in the borderlands and made broad use of his patronage network. There are numerous examples of him helping miners to find work on one or another of the properties he oversaw. He directed one correspondent to the manager at the New Cornelia Mine in Ajo for "rough work" underground, for instance, and got another former Rough Rider who had fallen on hard times a job as a chief watchman at the Copper Queen in Bisbee. Greenway also frequently helped relatives and friends of friends to get their start in the practical business of mining, "mucking underground."[6] Greenway's expansiveness and generosity toward miners whom he considered "good men," however, should not be misconstrued as a truly democratic sentiment, although it is possible that he considered himself egalitarian. Evidence of his alignment with company policies and principles is also easy to find, as Greenway was at the forefront of the infamous Bisbee deportation in 1917.[7]

By the late 1800s, due to changes within the mining industry and the profession of mine engineering, discussed in chapter 2, the technical skill of mining engineers was valued by mine promoters throughout the borderlands. With increased U.S. military presence through the Southwest after the Civil War, mining operations in Arizona, for instance, were becoming viable in ways they had not been earlier. As the odds that a given mining project would be successful improved, the drive on the part of the industry to obtain more capital also increased. Mining engineers remained an important piece of the fund-raising puzzle. Their expertise was still invoked as a reason to invest in a given mine, and mining engineers themselves, whether working as consultants, project engineers, or managers, often reported directly to a mine's investors either on paper or in person.[8] To be professionally successful, mining engineers depended upon their ability to merge the role of technical expert with other professional duties and obligations—namely, to reconcile their work as consultants and managers with different class, gender, and self-consciously "western" identities—as they asserted themselves within an industry that required a steady flow of capital. In the borderlands, this identity may have been borrowed explicitly from an earlier generation of managers and technical workers such as the notoriously crusty Ben Williams, an early manager at the Copper Queen in Bisbee.

As consultants, mining engineers had to communicate with mine owners, board members, and their agents—those people in the mining industry most associated with finance. As managers, mining engineers were responsible to the capitalists but principally interacted with mine superintendents, foremen, and miners in the field.[9] In addressing each audience, the engineers used distinct rhetorical strategies to present themselves as expert technologists who were qualified to assess mining claims and create plans for exploitation. Within their professional community, mining engineers adopted a discourse of identity that drew upon aspects of a distinctly American pioneer narrative to underscore their competence to assess and administer such risk.

From the 1860s onwards, the drama of working in western territories was a critical aspect of the professional image of mining engineers.[10] Raphael Pumpelly's memoirs exemplify the portrayal of mining engineers as men in the heart of "westering" conflict, although his tendency to portray himself as scared and overwhelmed is unusual for published work. Within a couple of decades, placing themselves in tales of the "Old West" had become one of the defining features of professionals working in this region. One engineer, who grew up in Missouri in a town whose bank was robbed in 1873 by Jesse James, was inspired, in part by this piece of local color, to study mining in hopes of seeing the "real" West. Of his first trip to New Mexico and Arizona, in 1888, he noted, "From what I read and had been told I expected the boys to wear their guns and spurs while dancing, but instead I found them a jovial lot."[11] An obituary for Louis Janin written by his old friend and colleague Rossiter Raymond made a great deal of Janin's early experiences with the Butterworth Expedition in Arizona, explaining that Janin's choice of career was directly related to the allure of the West he felt as a young man. Janin cared little for the domestic political and military dramas of the 1860s, according to Raymond. "[His] thoughts had long been turned to the new, wide, free region further West..., [he was] already enlisted for that war which...was waged by an army of prospectors and miners, for the physical conquest of a new Empire." Even James Douglas, mining engineer and head of the Phelps Dodge Company, which held mining operations throughout the borderlands, told stories about feeling like a tenderfoot when he first visited a bar in Arizona in 1881 and his fear at the time that someone would notice how foreign and threatening he found saloon culture.[12]

Mining engineers whose lived experiences of southwestern travel were legitimately "wild" did not necessarily enjoy those times in their careers when their lives were in danger.[13] In later years, however, those feelings neither prevented them from indulging in nostalgia for the "old west" nor stopped them from perpetuating those stories when they had the chance. The romance of the Wild West remained an important narrative strand in mining engineers' reported memories of their early years in the field. "I heard of Harry's adventures in Arizona [in 1865]," wrote Raphael Pumpelly to his friend Louis Janin about the latter's brother Henry, "and recognized the 'philosopher' in the coolness of his demand for spectacles, while under fire."[14] In this account, Pumpelly, who was familiar with how genuinely discomfiting western adventures could be, grants Henry Janin one of the most prized attributes of a man of learning: grace under fire. The "adventures" to which Pumpelly alludes occurred in Arizona, suggesting that Janin's party was besieged, possibly by the same forces that so traumatized Pumpelly. Pumpelly's admiration of Henry Janin's calm demeanor in requesting his glasses—something that clearly defined Janin as an educated man—indicates the significance these events held as touchstones within the profession. Here, Pumpelly suggests that even under extreme circumstances, a mining engineer thinks carefully and logically and demands the proper tools for pursuing his agenda.

The engineer as a westering adventurer was an image that inspired many young men at the start of their careers. As one mining engineer cogently recollected, he and his colleagues were "entranced and thrilled with...the glamour and adventure that appeared closely associated with such an occupation."[15] In letters home and in columns in local and national magazines, mining engineers portrayed themselves as quintessential American frontiersmen. Their chosen profession, they argued, "has a life, a speculation, a profit, [and] a vitality...well suited to the American character."[16] At the Colorado School of Mines, wearing a Stetson—the iconic hat of the western frontier—was a privilege reserved for "manly" seniors. Younger students eagerly awaited their chance to break in their hats.[17]

Many mining engineers explicitly related themselves and their working lives to the narrative of American westward movement.[18] This attitude toward mine engineering was summarized in a short essay entitled "The

Engineer Was Here," appended to a note thanking James Douglas for providing the author, Frank Aley, with transportation to Douglas, Arizona. While clearly a paean to Douglas personally, the qualities Aley attributed to Douglas were general rather than personal quirks. Aley describes Douglas as being like a "prosperous, intelligent farmer," necessarily attuned to detail and cautious by nature, but also like a railroad engineer, responsible for "pull[ing] the throttle that turns the main shaft" of the Phelps Dodge train. Of the "wonderful system of commercial progress in Arizona and Sonora... he is the Engineer."

In Aley's narrative of development, early mining engineers, including James Douglas, traveled to Arizona and Sonora when the country was wild and "pioneered" the establishment of mining camps and companies. When the companies prospered, mining engineers remained as the civilizing and productive forces in the economy, working, in Aley's words, "persistently, hopefully, and with profound discretion, playing the wonderful game of progression with Dame fortune." However overblown Aley's prose, he clearly links Douglas, and thus the mine engineering profession, to what historian Frederick Jackson Turner called the characteristic "American intellect": "acuteness and inquisitiveness; that practical, inventive turn of mind, quick to find expedients; that masterful grasp of material things... that restless, nervous energy."[19] Aley's many and conflicting analogies of a mining engineer's importance—the engineer as farmer, as driver, as economic mastermind—all relate to a concrete mechanical ability, the mining engineer's "masterful grasp of material things."

To Turner, the presumed individualism and intellectual attributes of the frontiersman were significant because they nurtured the growth of democratic institutions. For mining engineers, ingenious creativity in solving practical, technological problems was a key responsibility. This valorization of supposedly practical skills was internalized by mining engineers as an aspect of their own work out west. Indeed, a critical facet of mining engineers' identification with the pioneer spirit was that it served to protect them, at least in their own eyes, from accusations that their university training, designed to give a technical understanding of mining, actually rendered them soft and unable to understand the physical nature of the work.

As part of mining engineers' identification with a pioneering masculinity, they were careful to explain, both to one another and to their friends and families back east, that their work required physical stamina and tolerance of discomfort. Nobody, one engineer asserted, who lacked "the physique necessary to stand a great deal of hardship in all kinds of climate" should join the profession.[20] To reach the most remote mining camps in the Southwest and Mexico, engineers had to ride on horse or mule for days, sometimes weeks. The work of mapping and surveying a mine could take several months and involved walking miles a day, both above and below ground. As managers, good mining engineers spent several hours a day in the loud, hot, and stuffy world underground, actively supervising the work of miners, ensuring that cuts were made in the proper direction, or setting up and maintaining water pumps to clear the mine shaft. Engineers spoke among themselves disparagingly of "'the ladies walks' of the mine," which they had all at one time taken or led. The phrase described those mine tours that avoided all of the dirty or unpleasant places, suitable, therefore, for "ladies" or those men who needed to be "shielded" from the workings of the mine, such as eastern investors, or (probably more often) technical experts employed by rival mine owners.[21] As women were to be shielded from arduous or overly technical visits into the mine, so men who could not handle that rugged world were also to remain in a sphere wholly separated from the real work of mining. The implication is apparent that mining engineers themselves did physically and mentally challenging, thoroughly masculine work.

The primary audience for such demonstrations of masculinity was clearly other mining engineers, who liked to be reassured that they belonged in the field, and in their chosen profession, and that they were not "tenderfeet" with no practical knowledge of hardship. The ideology of manhood in the late nineteenth century is comprised of many strands of cultural meaning and is internally contradictory. Early mining engineers were mostly born to privilege, while those who got their start after the late nineteenth century were more likely to pursue the profession out of social ambition. Whether born to privilege or merely aspiring to it, mining engineers shared an interest with other privileged white men in projecting themselves as genteel, respectable, and of strong moral character—traits associated with a Victorian notion of manliness. Yet they also took pride in their physical endurance and the experiences that set them apart, which called for a more modern concept of masculinity, one emphasizing physical strength, risk taking, and virility.[22] In this emphasis on physicality, they followed the very public example of Theodore Roosevelt, who valorized the outdoors and held himself up as an embodiment of the strenuous life.[23] Indeed, as the scion of an elite New York family who traveled west to improve his health but discovered the glory of the American frontier, Roosevelt's experiences and worldview resonated with many mining engineers, whose personal and professional trajectories mirrored the president's. At a time when professional men chose safe, sedentary, interior livelihoods, mining engineers were engaged in a physically demanding profession that had them battling the elements more often than not.[24] Further, in contrast to their peers in other realms of engineering who did not tend to work in remote locations or operate under such physical hardship and generally spent a great deal more time in the drafting room, many mining engineers loved outdoor life. They gloried in beautiful vistas when traveling to mining sites and considered their work respectful of nature's bounty rather than its violent extraction. Indeed, many considered working in the rugged wild to be a chief benefit of their line of work.[25]

The desire of mining engineers to embrace a more virile concept of manhood was apparent into the twentieth century, when mine investors were more likely to travel to mining sites to see the prospects for themselves. By this time, such travel was no longer quite as daunting as it had been in earlier decades, although it still had its challenges. The physical risks that earlier regional mining engineers took as a matter of course simply to reach mine sites were no longer necessarily a part of the job. Rather, mining engineers now supported their adopted identity as virile outdoorsmen by acerbically distinguishing the discomforts inherent to their work from the comforts embraced by their more sedentary counterparts. Joseph Obermuller smugly recounted meeting a group of investors near Nacozari, Sonora, about 1905. They were unprepared for the difficulties of travel in Mexico and had been obliged to reach the Tigre Mine on mules.

> Anyone that ever rode a freight mule can best understand what these men must have endured, especially as most of them had never been in a saddle—all being business, or one should say office, men not used to hardships.... Some of them were sore in mind and body. Others got a thrill and were inclined to joke, while others resented anything funny.[26]

This account also implies that Obermuller himself was no stranger to the discomfort of riding a freight mule, and, by extension, that this was something most mining engineers would have experienced at one time or another. Tales of the antics of investors in the field and the utter lack of physical confidence demonstrated by so-called "office" or "business" workers were a trope of mining engineers' storytelling. Such mockery asserted the engineers' physical superiority over their employers. But it also served a parallel purpose to the mockery of mining engineers by working miners, a comfort

Mining engineers in the field, 1910. They are obviously posing for the camera, showing off their instruments and also the engineer's notorious, tall laced boots. Courtesy of the University of Wyoming, American Heritage Center, Samuel H. Knight Papers, 172333.

to men with little control over the vagaries of their existences, at least as related to the mining industry.[27]

With such a clear sense of themselves as physically tough, mining engineers were particularly sensitive to the charge that they were unable to understand the needs of ordinary miners. Their attempts to manage a labor force were intimately connected to their understanding that they were sophisticated professionals with a distinctive western, and thoroughly masculine, status.[28] In management, mining engineers embraced the suggestion that they were rugged western adventurers working in one of the most physically demanding industries in nineteenth-century America. Their imagined identity as frontiersmen had particular resonance in the borderlands region in the later decades of the nineteenth century, ultimately influencing their work as supervisors in the homosocial labor world of the mine.

Yet, there is an internal contradiction in mining engineers' embrace of the frontiersman iconography. An anti-intellectual bias is evident when presenting themselves as such, which is in notable contrast to their actual class background and antithetical to what mining engineers are by definition: university-educated professionals. Historian Clark Spence described western mining engineers in the late 1870s and 1880s as "remarkably sophisticated men...no other group in the West was as well traveled and as well educated."[29] In fact, by virtue of their formal education and generally broad experience in the world, mining engineers were members of the late-nineteenth-century cultural elite. The erudition of educated mining engineers served them well in the drawing rooms and social clubs of New York or San Francisco, where many maintained permanent offices. But this same university training, which assured mining engineers of their status as technical experts and provided the social authority for their work, also implied a lack of hands-on experience that mining engineers strove to combat.

Agreeing to take control of the day-to-day workings of a mine was a calculated risk, one that could end poorly for engineer and mine owner alike. Experienced mining engineers were well aware that the culture of apprenticeship, which dominated western mining labor into the late nineteenth century, valued length of service over any depth of theoretical knowledge. Although the complexity of hard-rock mining demanded the managerial and technical expertise of university-trained mining engineers, any operation that sought high-grade ore—which is to say, most heavy-metals mining prior to the turn of the century—also relied on skilled miners, men who knew first hand when an engineer made a bad decision, as they were the ones searching for nonexistent ore bodies.[30] As one engineer explains, a skilled miner could "find…out very quick whether his superior possesses the required knowledge, or not, and if not, is impudent enough to criticize and ridicule him."[31]

Exacerbating the anxiety many mining engineers felt about exerting their authority over skilled miners in remote locations was their deeply conflicted sense of masculine identity.[32] Historian of engineering Ruth Oldenziel discusses the role of class in the dueling notions of manhood that were present in canal-, rail-, and bridge-building labor camps of the late nineteenth century. Oldenziel argues that such "labor camps…were largely societies of men, where hard living, hard working, and hard drinking were cherished values, reminding many an aspiring engineer of the kind of proletarian manhood they were determined to avoid at all costs."[33] Aspiring civil and mechanical engineers worked cheek-by-jowl on these construction projects with skilled and unskilled immigrant laborers from around the world. These engineers set themselves apart from the mass of manhood surrounding them through a variety of markers. They considered working as a laborer to be a "productive" form of physical exercise, but Oldenziel notes that although aspiring engineers saw in the carousing and raucous society of labor camps a form of working-class manhood that they sought to avoid, they were also drawn to the proclamations of brotherhood and solidarity that formed the backbone of working men's culture in these places. The disjuncture in these images of masculinity, Oldenziel argues, could only be overcome by seeking a professional career—one that valorized practicality, physical fitness, and the fraternal world of working men—all of which were component parts of the budding profession of engineering.[34]

Mining camps were very much like the work camps Oldenziel describes: male-dominated societies, replete with opportunities for men to prove themselves in contests and competition. Yet unlike the civil and mechanical engineers of whom Oldenziel writes, it is not clear that mining engineers were terribly concerned with setting themselves apart from the performance of manhood as enacted in mining camps, although they were instantly recognizable on site by their sartorial choices. Mining engineers favored wide-brimmed hats, tailored shirts, jackets, trousers, and tall laced boots. They placed much greater emphasis on the physicality of their work than did civil or mechanical engineers; they were not front-office workers. They used their identification with "pioneering" masculinity not only to counter their own discomfort with the more proletarian masculinity of working miners but also as a way to understand ordinary miners so as to inform their own labor management tactics.

While mining engineers adopted a discourse of physical valor when interacting with their social peers, they emphasized other aspects of their professional identity when interacting with miners, particularly when working as managers and in circumstances in which their role was to represent the interests of a mining company. For instance, mining engineers agreed that it was important to spend time with miners, not to learn about their work, but to get to know them as men. "The personal equation is everything," one engineer stated, because "when you get a large number of men under one management you obliterate that personal equation and so render the relations unhuman."[35] It is evident, however, that mining engineers generally did not socialize with ordinary miners, nor for the most part did they have much empathy for the plight of the working man, so this exhortation to get to know miners as individuals should not be taken literally. Rather, it speaks to a desire to keep relations on a human scale such that mining engineers and workers could interact "man to man." One engineer wrote a treatise for an investor on the skills needed to manage a mine in Mexico. Chief among them was the exhortation to "be just towards the common men" in order to "be sure that the employees will work hand in hand with him, as they look at him as their superior…in knowledge."[36] The ability to feel compassion and to behave with fairness toward other people was a central tenet of middle-class manhood in the late nineteenth century, and mining engineers evidently embraced this ideology wholeheartedly.[37]

Such rhetorical focus on the importance of "man-to-man" interaction between mining engineers and workers was perhaps embraced by engineers to convince themselves that their opposition to labor unions was due to disdain for the tactics, rather than the goals, of organized labor. After all, the chief stated goal of an organization such as the Western Federation of Miners (WFM) was to improve the health and well-being of miners, a project that directly benefited the work of mining engineers who were well aware that a contented workforce was better than a dissatisfied one. The awkward fact that WFM membership was restricted to white men only, when the population of mine workers in the borderlands was racially and ethnically diverse, was never (to my knowledge) commented on, and perhaps went unnoticed by area mining engineers.

Contented miners produced quality mining work and had lower-than-average turnover, which could be as high as one-third of the labor force each month in Arizona at the turn of the century.[38] The reasons mining engineers opposed unions are relatively self-evident. Chief among a mining engineer's tasks was to improve the efficiency of a mine: cut expenses and raise production. By agitating for higher wages, labor unions undercut the first of these.[39] Strikes, of course, were a particularly powerful tool for union members and organizers, and such industrial actions cut at the heart of the work of mining engineers by destroying carefully calculated estimates of expenses and production. In their opposition to unions, therefore, mining engineers embraced that aspect of their professional identity noted by Edwin Layton—their professional self-interest aligned with the demands of capital.

When the Western Federation of Miners tried to unionize workers in the Southwest, for instance, many mining engineers were infuriated, considering that the union created problems rather than solving them. "It was a fairly contented and happy community," William F. Staunton wrote of the Congress Mine in Prescott, Arizona, "until the Western Federation of Miners began to try to unionize it." According to Staunton, even in its formal petition to open shop, the WFM issued "no complaint whatsoever about wages, working or living conditions, which were specifically stated to be satisfactory." He writes that shortly afterward there was a strike necessitating a mine shutdown, and pictures of Staunton's foreman appeared in the paper under the heading "King Scab in Arizona."[40] Another managing engineer, after experiencing threats on his life and uncovering a plot to dynamite his house, considered his best weapon against the WFM to be what he deemed honest and open communication with the working miners. He instituted a policy of only hiring workers who could be eligible for union membership—in other words, only white men—yet never explicitly forbade them from joining the union. He believed that by offering them what he called "a day's work for a day's pay" he could undercut their desire to join the WFM. This strategy was fairly common in Arizona. Indeed, his ostensible argument with the union was not that it existed but that it demanded a closed shop, which he considered un-American.[41] However, the evident vitriol that at least some miners felt toward this engineer belies his belief in the power of man-to-man interaction.

In addition to the knowledge claims that mining engineers were able to make if they had "practical" experience as a miner, they also extolled such experience as a way to improve one's sense that he had the ability to speak to miners effectively.[42] L. D. Ricketts explained to an interviewer in the 1920s,

> Manual work gives... [a mining engineer] the opportunity to know the viewpoint of the workman; and, if in the future he is called upon to handle men, such an experience is immensely valuable to him. This is merely an opinion, because I have not worked either as a miner or a smelter-hand; but I have always regretted that I did not have a little experience of the kind, in order that I might be closer to the worker's viewpoint.[43]

When he made those remarks, Ricketts was superintendent of one of the largest copper mines in the western hemisphere. He was known not only for his tremendous success building southwestern mines but also for what might be described as an ostentatious modesty. He affected rumpled, work-stained attire, scuffed boots, and worn headgear—a dress code greatly at odds with his background as a Princeton PhD and his influential work as chief engineer for Phelps Dodge. Peers described him as "common as an old shirt" and "not much on clothes." Despite, or perhaps because of, Ricketts's self-conscious realization that he did not fully understand the physicality of mining, throughout his career he was known for shunning many external markers of success. In his professional work, he was well-known as a generous consultant, always seeking the opinion of others. In the words of the perennial commentator on mining engineers, T. A. Rickard, editor of the *Engineering and Mining Journal*, Ricketts was "not cocksure, but deliberate." Stories abounded in mining communities about people who mistook him variously for a miner, a vendor, or even a hobo.[44] But in fact his personal appearance garnered so much attention precisely because it was so at odds with what was normal among his fraternity.

Ricketts's stated regret that he was unable to legitimately empathize with his workers is further evidence that mining engineers had many reasons to embrace the more rugged physicality of miners rather than flaunt their own social status. Such an assertion can readily be interpreted as a gloss of Ricketts's desire to "understand" working miners in order to manipulate them, or, in a more sinister fashion, subvert their attempts to organize for themselves. To some extent, of course, this is true: mining engineers did not like working with unions, which they rightly understood to be powerful political forces that could easily disrupt mining operations and destroy hairline budgets. Yet Ricketts was not completely cynical. The prevalence of a discourse of identification with working miners in trade publications as well as personal letters and memoirs; the critical importance of programs such as the Columbia Summer School to get budding mining engineers into the tunnels; and the continued need for skilled, apprentice-trained miners through the early twentieth century all point to real advantages for mining engineers who could easily grasp how their directives sounded to a worker. An experienced miner was likely to have spent time working with both good mining engineers and poor ones and to have worked his share of successful and unsuccessful mines. He could tell when a mining plan might fail, because he had seen it happen before.

The rationale behind the desire to empathize with working miners has been overlooked by historians of labor, who tend to assume that as mining engineers were structurally opposed to mine workers in labor disputes, they operated from a position of cynical contempt. Some scholars suggest that mining engineers used their "expert knowledge" to "gain leverage over the workers" in Arizona toward the end of the nineteenth century.[45] But what leverage, precisely, did highly educated engineers have over the average mine worker that less well-educated, apprentice-trained, "up through the ranks" foremen and superintendents did not? Working miners were not the audience for mining engineers' displays of erudition; their supervisors were. Whether miners felt any kinship with engineers who had experience with manual labor is somewhat ambiguous, although it is likely that such experience could generate respect. As Frank Crampton, a miner-turned-mining engineer, observed of his fellow engineers, "Most of them were too far up in the clouds to have truck with ordinary hard-rock stiffs. They would have done better and learned more had they comedown [*sic*] the earth."[46] By this, Crampton meant that the miners thought engineers did not understand how mining labor was organized or the tremendous physical labor it took to extract ore from the ground.

Another common interpretation of the rift between mining engineers and working miners is that the introduction of mechanized technology into western mining camps, attributable to the presence of engineers, led to the deskilling and proletarianization of miners. In turn, deskilling led to an increasing radicalization of the mining labor force. This interpretation assumes engineers were perpetually in opposition to working men.[47] Nineteenth-century mining engineers, however, were invested in the lives and experiences of both labor and management, and the pride many mining engineers felt in the physicality of mining work was useful for them in negotiating that liminal position.[48] Although many engineers did serve as the de facto face of the company in on-the-ground negotiations with working miners, they tended to bemoan labor conflicts, not only because they lost money but because they gave voice to the inequities of mining camps that engineers, with their valorization of physical work and pioneering identity, would likely have preferred to ignore. Trained to read geographic landscapes, to extract ore from hard rock, and to build complex machinery, mining engineers were poorly prepared by their education to handle labor issues. Courses in administration and accounting were not typically added to university programs until the twentieth century and even then took a back-seat to math, physics, and mechanics.[49] In fact, when mining engineers expressed strong feelings about labor issues, their voices were likely to be ignored.[50] Thus, they had a strong incentive to focus on technical rather than personnel issues, and the evidence is strong that they weighed in on the actual question of which workers to hire, and how to treat them, only under duress.

A salient aspect of the class distinction in borderland mining camps that mining engineers ignored rather more successfully were the racial and ethnic divisions within mining camps. In general, mining engineers appeared to have shared in the anti-Mexican sentiment common in the United States in the nineteenth century. These beliefs consisted of thoughtless stereotyping of Mexicans as lazy, dirty, and unethical, and Mexican food as gruesome. There was also a general sense that Mexican mining was characterized by "indifference, waste and insufficient machinery."[51] Given how closely many mining engineers worked with non-Anglo workers, however, it might be expected that they could be clear-eyed about the antagonism bred by such attitudes. Raphael Pumpelly, for instance, once observed that the Mexican workers "felt only hatred" toward their American employers in the silver mines near the border. He acknowledged that this may have had something to do with the American companies' practice of "paying the Mexicans the greater part of their wages in cotton and goods, on which the company made a profit of from one hundred to three hundred per cent"—hardly a circumstance to warm the relationship between employer and employed.[52]

Pumpelly, as usual, was remarkably farsighted, compared to many of his colleagues. His friend and classmate Louis Janin, by contrast, was bullish on the subject of a company store and considered it a significant source of operating revenue at any mine site, since paying miners in scrip led to a "double gain" for the company—profit on the labor or the miner and worker debt at the company store.[53] A common sentiment among mining engineers was that "Mexican labor is cheap, and if properly managed can be used to great advantage," yet always needed to be trained to accomplish tasks that white Americans were already competent to fulfill.[54] That sentiment, readily recognizable to a twenty-first century observer as racist, was premised upon an obvious structural condition of work at the mines: many white mine workers were hired from the outside as skilled workers—blacksmiths, furnacemen, or mechanics. Any locally available unskilled workers, Mexican or otherwise, would have to be trained to fill such specialized positions.

What Pumpelly observed of the poor treatment of Mexican (and later, Mexican-American) workers by American companies was only the beginning of a clear practice of discrimination, one that mining engineers, as a group, failed to protest: the dual-wage system. The practice of one pay scale for white workers and another for nonwhites was widespread, justified by the prevalent belief among mining engineers that nonwhite workers were more ignorant than their white counterparts. One mining engineer in southern Arizona summarized the attitude toward the available workers when he fitfully complained,

> We have found it exceedingly difficult to get competent labor of all classes.... We find that the Mexicans we get here are of a very disreputable and tough class, and on pay days generally cause riotous disturbances. We, however, employ them where skill is not required....[55]

A particularly striking example of the income disparity was in place at the Longfellow Mine in 1880, where Chinese miners were paid $40 per month, Mexican miners $50 per month, and "American" [white] miners $75 per month.[56] The Chinese and Mexican workers were underground ore miners; the Americans were on "dead-work," or work that was not directly profit-making, such as preparatory work for opening a mine and, later, shoring up mine shafts and routine maintenance. Although dead work is critical to the functioning of a mine, it was often underestimated by mine owners and operators. Indeed, the cost of dead work was often estimated by mining engineers at approximately $1 per ton of ore produced, while ore extraction cost three times as much.[57] That the Longfellow management entrusted the usually higher-valued underground work to nonwhite workers indicates that the company was probably in dire financial straits. The ingrained nature of the dual-wage system is underscored at a mine where white workers, doing work deemed of secondary importance, were paid almost double the wages of nonwhites performing critical preparations and maintenance.[58]

Despite mining engineers' reluctance to put Mexican workers in positions of responsibility, many engineers took small steps to make those workers on whom a mine was dependent more comfortable. As managers, engineers consistently looked for ways to encourage worker productivity or to head off any efforts to unionize, which many engineers considered to be the same thing. In 1889, for instance, William Church established both English- and Spanish-language reading rooms for employees at the Phelps Dodge properties in Morenci, Arizona. He believed that access to these small libraries was critical to maintaining the morale of the workers and wrote to his principals requesting reading materials.[59] As early as 1885, some mining engineers in Mexico advocated hiring men for three eight-hour shifts, rather than for two ten-hour shifts, on the understanding that underground work was simply too exhausting to be undertaken for ten hours straight.[60] The rumor circulated in 1900 that the United Verde Mine at Jerome raised wages 15 percent and put men on an eight-hour shift. In response, an engineer at the nearby Ray Consolidated Mine told his board of directors, "This is liable to cause trouble to other mine owners," and, of course, to other engineers.[61] Engineer John G. Greenway, who as the president of the Calumet and Arizona was not known for generous policies toward nonwhite workers, established Sunday as a day of rest throughout his mines in Arizona, a progressive action for 1910.[62] Greenway's decision may well have been made for the most cynical of reasons—to undercut a strike effort and to police the leisure-time activities of the company's workers, but as a mining engineer who had spent considerable time underground he believed that a mandated day of rest materially benefited the lives of workers at the C & A, making them more effective laborers.

Although there is no evidence that engineers colluded, or even formally discussed, wage scales or welfare provisions for working miners, the contradiction within their labor-management goals—the need to placate workers and the need to keep labor costs low—was difficult to overcome. Mining engineers had a practical desire to stave off unionization without resorting to violence or other disruptive measures. The rhetorical emphasis on middle-class decency was a weak attempt to assert a consistent labor strategy over an aspect of mine engineering work that fell outside the parameters of technical knowledge yet was not as susceptible to manipulation via the vigorous assertion of a frontier masculinity as were mining engineers' peer-to-peer relations.

Among themselves, and to a lesser extent in the public sphere of the eastern United States, mining engineers affected a distinctly western masculinity, one that allowed them to feel comfortable among the men they supervised and interacted with in the field. Nothing could be more damaging to a young engineer's standing at a mine than to be labeled a tenderfoot, and they worked hard to escape that designation. Mining engineers gave themselves a means to empathize with ordinary miners. In turn, this identification with miners as *men* provided mining engineers with a primary way to understand how to manage workforce relations and to understand what it meant to be a "working stiff" in a hard-rock mining camp.

During the 1880s through roughly 1910, the most aggressive era of U.S. capital investment in Mexican and southwestern mining projects, mining engineers negotiated a set of overlapping gender and class identities to legitimate themselves to an audience of employers, employees, and other mining engineers. An engineer's ability to create a real or imagined bond with the men he supervised or directed could make or break a mining enterprise. In the field, the popular image of mining engineers as physically adventurous men was an asset for mining engineers. Their elite status, however, could stand in the way of his relationship to ordinary miners because it engendered resentment and failed to provide engineers with the tools they needed to be successful managers of labor. Invoking western and masculine identities helped borderland mining engineers overcome this "deficiency" in their training. Asserting a blandly middle-class morality, with an emphasis on the importance of "man-to-man" interaction, could help mining engineers elide their own discomfort with the class divisions within a mining community.

Notes

1. Mary Hallock Foote, in Paul, ed., *A Victorian Gentlewoman*, 133, 130. Arthur Foote eventually found steady employment working with his brother-in-law, James D. Hague, at Grass Valley, California.

2. Richard H. Peterson demonstrates that although men who were defined as

"mining magnates" experienced tremendous social mobility on the mining frontier between 1870 and 1900, they were exceptional in their communities, which were characterized by a racially and ethnically diverse, economically impoverished, itinerant workforce; Peterson, "The Frontier Thesis and Social Mobility on the Mining Frontier," *Pacific Historical Review* 44, no. 1 (1975):52–67. Wallace Stegner's Pulitzer Prize-winning *Angle of Repose*, loosely based on the lives of Mary Hallock Foote and her husband Arthur D. Foote, offers a lightly fictionalized portrayal of the intermingling of mining engineers and elite New York society in the 1870s.

3. Layton, *Revolt of the Engineers*, 33, 35. Ochs, "The Rise of American Mining Engineers," offers a detailed analysis of the work undertaken by graduates of the Colorado School of Mines in its first years of operation and argues that although there appears to be a clear distinction between the managerial and the technical tasks that mining engineers were asked to undertake, and that the educational system was structured around this distinction, in practice, nineteenth-century engineers tended to hold positions that required them to be both managers and technical workers. See Ochs, 283–284, n. 13–14.

4. Although a few well-known mining engineers such as James Douglas spent the bulk of their careers working for only one company, in the nineteenth century the vast majority of mining engineers changed employers frequently.

5. John Greenway to Theodeore Roosevelt, April 20, 1911, and John Greenway to Theodore Roosevelt, December 3, 1912, folder 2718, box 192, Greenway Collection. Greenway is also held up as a pioneer in public memory in Arizona, as indicated by the prominent display of his statue outside of the Arizona Historical Society in Tucson. He appears to have changed his mind about investing abroad later in life. After 1919, Greenway held a one-third interest in the Ahumada lead mine in Chihuahua, where he served as an absentee general manager.

6. John Greenway to Theodore Roosevelt, December 28, 1914, folder 2718, box 192; Annesley Young to John Greenway, February 6, 1914, John Greenway to Annesley Young, February 13, 1914, folder 2722, box 192; John Greenway to Frank Smith, February 7, 1913, folder 2721, box 196, Greenway Collection.

7. John Greenway to Theodore Roosevelt, August 9, 1910, box 192. The Bisbee deportation lies outside the bounds of this study, but there is a fascinating archive detailing Greenway's involvement in folder 2393, box 181, Greenway Collection. For more on this subject see Benton-Cohen, *Borderline Americans*, ch. 7.

8. Michael N. Greeley, "The Early Influence of Mining in Arizona," in Greeley and J. Michael Canty, eds., *History of Mining in Arizona*, vol. 1 (Tucson: Mining Club of the Southwest Foundation, 1987), 18–22; Robbins, *Colony and Empire*, 25; Buder, *Capitalizing on Change*, 108–109; 123–125.

9. Mining statistics of any kind are difficult to obtain for the 1860s–1880s. By the 1890s, Clark Spence estimates that approximately one-third of university-trained mining engineers were working in managerial positions. Spence, *Mining Engineers*, 139.

10. On American companies mining in the borderlands in the 1850s and 1860s, see chapter 1. Pumpelly, *Across America and Asia* and *My Reminiscences* (New York: Holt, 1918). Other widely read portrayals of mining in Arizona include Browne, *Adventures in the Apache Country*, and Charles Poston's narrative essays, collected in *Building a State in Apache Land: The Story of Arizona's Founding as Told by Arizona's Founder*, ed. John Myers (Tempe, AZ: Aztec Press, 1963).

11. "Life and Travels of a Mining Engineer," folder 2, Joseph E. Obermuller Papers, Arizona Historical Society, Tucson, AZ (hereafter JEO). For examples of the discussion of native hostilities, see E. E. Olcott to W. E. Dodge, September 20, 1881, and W. E. Dodge to James Douglas, October 12, 1881, folder 30, box 3, James Douglas Papers, Arizona Historical Society, Tucson, AZ (hereafter JD).

12. Rossiter W. Raymond, [Louis Janin Obituary], *AIME Transactions* 49 (1914):831–836; [Introduction to the West], folder 53, box 4, JD. For more on Pumpelly, see chapter 1.

13. See chapter 1.

14. Raphael Pumpelly to Louis Janin, 1865, box 5, Pumpelly Papers II.

15. Fred Bailey, no. 115, Oral History Collection, C. L. Sonnichsen Special Collections, University of Texas, El Paso, TX.

16. Almarin B. Paul, "Gold and Silver Mining," *Pacific Coast Annual Mining Review and Stock Ledger...* (San Francisco: Francis and Valentine, 1878), 129.

17. LeCain, *Mass Destruction*, 57.

18. "Life and Travels of a Mining Engineer," folder 2, JEO; Frank Aley, "The Engineer Was Here," folder 77, box 5, JD, AHS.

19. Frederick Jackson Turner, "The Significance of the Frontier in American History," 37.

20. T. A. Rickard, *Interviews with Mining Engineers* (San Francisco: Mining and Scientific Press, 1922), 111.

21. [JEH] to H. M. Dieffenbach, Esq., Mexico City, June 29, 1907, folder 53, box 3, Hyslop-Beckmann Collection, University of Texas, El Paso, TX.

22. Gail Bederman, *Manliness and Civilization: A Cultural History of Gender and Race in the United States, 1880–1917* (Chicago: University of Chicago Press, 1995), 7, 10–11, 18–19.

23. Bederman connects T. R.'s embrace of an identity as a "cowboy-fighter" to the racial imperialism of the westward expansion of the United States and to Roosevelt's own concern about the decline of the white race. See Bederman, *Manliness and Civilization*, ch. 5. "The Strenuous Life" is the title of a speech Roosevelt gave in 1899 setting out his philosophy that vigorous activity was necessary for the health and well-being of the nation. Although Roosevelt's purpose in the speech was to argue for a vigorous and engaged foreign policy—in particular, war with Spain and the construction of the Panama Canal—a secondary thrust of the speech was in support of individual action: "The highest form of success [comes]...to the man who does not shrink from danger, from hardship, or from bitter toil, and who out of these wins the splendid ultimate triumph." Theodore Roosevelt, *The Works of Theodore Roosevelt* (New York: Century, 1901), 3–22.

24. Anthony Rotundo, *American Manhood: Transformations in American Masculinity from the Revolution to the Modern Era* (New York: Basic Books, 1993), 167–168.

25. LeCain, *Mass Destruction*, 55–57.

26. "Life and Travels of a Mining Engineer," folder 2, JEO.

27. Many scholars have pointed to a sense of independence as the crucial identifying attribute of self-consciously "western" men. See Laura McCall, "Introduction," to Matthew Basso, Laura McCall, and Dee Garceau, *Across the Great Divide: Cultures of Manhood in the American West* (New York: Routledge, 2001), 3. On the mockery of mining engineers by working miners, see chapter 2.

28. Amy Greenberg's concept of "aggressive" antebellum manhood is a better

model for "western" masculinity than a more amorphous "beholden to no man" sentiment. Greenberg, *Manifest Manhood*, 13.

29. Spence, *Mining Engineers*, 332; John B. Rae, "Engineers are People," *Technology and Culture* 16, no. 3 (1975):414.

30. Hovis and Mouat, "Miners," 434–435; 440–441.

31. Spence, *Mining Engineers*, 239; Ottokar Hoffmann to George Crane, Esq., November 5, 1885, K-3, JDH.

32. A discussion of the tension within the masculine identity of professional men during the Gilded Age can be found in Rotundo, *American Manhood*, esp. chs. 8–9.

33. Oldenziel, *Making Technology Masculine*, 55.

34. Oldenziel, *Making Technology Masculine*, 57–58.

35. Rickard, *Interviews*, 184.

36. Ottokar Hoffmann to George F. Crane, November 5, 1885, K-3, JDH.

37. Michael Grossberg, "Institutionalizing Masculinity: The Law as a Masculine Profession," in Mark C. Carnes and Clyde Griffen, eds., *Meanings for Manhood: Constructions of Masculinity in Victorian America* (Chicago: University of Chicago Press, 1990), 144–146.

38. Huginnie, "'Strikitos,'" 141; Hovis and Mouat, "Miners," note 65, 452.

39. Larry J. Griffin, Michael E. Wallace, and Beth A. Rubin, "Capitalist Resistance to the Organization of Labor before the New Deal: Why? How? Success?" *American Sociological Review* 51, no. 2 (1986):149–150.

40. Staunton, "Memoirs," 79, 121–122, box 1, WFS.

41. Courtenay DeKalb to James Douglas, October 29, 1904, folder 6, box 12, AZ 290, Lewis Douglas Papers, University of Arizona Libraries, Special Collections, Tucson, AZ (hereafter LD).

42. See chapter 2 for an extended discussion of "practical" education. Rickard, *Interviews*, 435; Packard, "Mining Schools," 173–174; Balch, *Mines*, 271–344; [Morris Parker Memoir], 45–46, 54, folder 3, box 1, Papers of Morris B. Parker, The Huntington Library, San Marino, CA (hereafter MBP).

43. Rickard, *Interviews*, 435. Further discussion of Ricketts's work in the copper industry can be found in chapter 5.

44. Walter R. Bimson, *Louis D. Ricketts (1859–1940): Mining Engineer, Geologist, Banker, Industrialist, and Builder of Arizona* (New York, NY: Newcomen Society of England, American Branch, 1949), 19; T. A. Rickard, *Interviews*, 455.

45. Spence, *Mining Engineers*, 176–178; Huginnie, "'Strikitos,'" 124.

46. Frank A. Crampton, *Deep Enough: A Working Stiff in the Western Mine Camps* (Denver: Sage Books, 1956), 51.

47. Wyman, *Hard Rock Epic*; Brown, *Hard Rock Miners*. Hovis and Mouat persuasively argue that Wyman and Brown romanticize the hand miners and thus blame labor deskilling on the process of mechanization—the introduction of machine drills and the use of explosives. While these tools were dangerous and undoubtedly contributed to labor unrest, Hovis and Mouat point to the rise of nonselective mining, or mass mining, as the primary cause of the deskilling of mine labor in favor of the expertise of engineers. Hovis and Mouat, "Miners," 429, 439, 332–449.

48. Oldenziel, *Making Technology Masculine*, 74–75.

49. Governing Board Minutes, February 5, 1906, folder 56 box 6, RU 819.

50. Sarah E. M. Grossman, "Capital Mediators: American Mining Engineers in the U.S. Southwest and Mexico, 1850–1910" (University of New Mexico, PhD diss., 2012), ch. 3.

51. See for instance a characterization of Mexico as a land of "frijoles and . . . tortillas, patted by the unwashed hands of dirty Mexican women" in Edward Reilly to James Douglas, February 14, 1881, folder 29, box 3, JD; Louis Janin Jr. to James D. Hague, July 12, 1889, M -15, JDH; [unknown], "Preliminary Report on the Smelter at Bonanza, Zac., Mexico, and on some of the Mines in that Vicinity," December 24, 1901, folder 4, box 6, Harold A. Titcomb Collection, American Heritage Center, University of Wyoming, Laramie (hereafter HAT); *Prospectus of the Loma de Platta, or Hill of Silver Mine, of Mexico, Situated in Altar District, State of Sonora, Republic of Mexico* (Atlantic City, NJ: Telegraph Steam Printing House, 1880), 5.

52. Pumpelly, *Across America and Asia*, 32.

53. Diary 1863, HM 64294, and Diary 1863–64, HM 64295, Louis Janin Papers (Addenda).

54. Ottokar Hoffmann to George Crane, Esq., November 5, 1885, K-3, JDH.

55. "Report of the General Manager of the Ray Copper Mines . . ." 30 June 1900, P108, JDH.

56. "Extract from the Report of Arthur W. Wendt on the Longfellow Mine," January 1880, folder 91, box 5, JD.

57. Joseph Henry Collins, *Principles of Metal Mining* (New York: G. P. Putnam's Sons, 1874), 34, 65; The *Engineering and Mining Journal* bemoaned the fact that mining engineers tended to underestimate the expense of dead work when writing up their reports on mines. "It is dangerous not to make liberal allowances for sums not needed directly in the extraction of ore. . . . How many promising mines have been swamped during the past few years because disappointment followed in wake of 'extraordinary expenses'?" *Engineering and Mining Journal* (March 1, 1884):153.

58. Except for Raphael Pumpelly's observation about Mexican workers in Tubac in the 1860s, cited above, I have seen no evidence that any mining engineers objected to the dual-wage system or advocated for higher wages or better housing conditions for working miners.

59. William Church to William Dodge, May 7, 1889, folder 4, box 2, LD.

60. [Morris Parker Memoir], 96, folder 1, box 1, MBP; Ottokar Hoffmann to George Crane, Esq., November 5, 1885, K-3, JDH.

61. "Report of the General Manager of the Ray Copper Mines . . . ," June 30, 1900, P108 JDH.

62. Benton-Cohen, *Borderline Americans*, 130; John Greenway to Theodore Roosevelt, August 9, 1910, folder 2718, Greenway Collection.

CHAPTER FOUR

RHETORIC AND RISK

The Performance of Objectivity at the Copper Queen Mine

In the mid-nineteenth century, most mining engineers who worked in western North America were employed as consulting engineers. Hired to inspect and report on mines and mining prospects, these men were tasked with going into the field as the eyes and ears of hopeful investors, using their professional expertise to construct a report to give employers a sense of whether a mine was a "good prospect" or not. Engineers usually discussed this work as if it were a straightforward proposition. A technical expert, after all, might be expected to provide an objective report on the reality of a mining prospect, but it was actually a rather complicated endeavor. Under the best circumstances, a mining engineer had to balance his own findings and opinion about the mine site with his knowledge of the financial constraints and ambitions of his employer. Under less-than-ideal circumstances—and mining engineers operated under less-than-ideal circumstances most of the time—the engineer had to wade through contradictory and obscure evidence on site while constructing a report for his employer that appeared to be objective and straightforward.

A mine report is a study in situated objectivity. Mining engineers were tasked with giving their expert opinions in the midst of a variety of interested parties whose agendas did not always align and whose needs did not always fit the evidence a mining engineer might find. The earliest formal mining reports for the Copper Queen Mine in Bisbee, Arizona, are one by John Daniell, an engineer whose principal work was at the Lake Superior copper mines in Michigan, and one by James Douglas, who went on to run the Copper Queen and then to be president of Phelps Dodge. The reports exemplify the challenges that mining engineers faced as they sought to offer

Copper and Silver borderlands, 1890s. Map by Bill Nelson.

clinical and objective analyses of mine sites, while underscoring the ways in which they used their expert knowledge to guide the decisions of investors and financiers.

The Early History of the Copper Queen

In the 1850s and 1860s, Americans who traveled into the Southwest were drawn by the promise of precious metals. Silver mines in places such as Tubac, Arizona; Parral, Chihuahua; and Cucurpe, Sonora, showed the promise of good returns. Mining engineers such as Raphael Pumpelly, Guido Küstel, and Louis Janin, touring and surveying the region during the 1850s and early 1860s, provided potential investors with a wealth of information that emphasized the native value of local mines. But they also cautioned that transportation, logistics, and general lack of security throughout the border territories meant returns on investment in precious metals mining would be modest at best.[1] Such relatively sedate assessments of the region's mineral wealth, however, did little to dissuade prospectors from scouting the landscape for likely outcroppings. By the 1870s, much of central Arizona was given over to silver mining, and the Arizona Territory's status as a prime destination for prospectors was well established.[2]

One such prospector was Ed Schiefflin, who did some scouting from the army base at Fort Huachuca. In 1877, Schiefflin staked a silver claim in the far southeastern corner of Arizona. He named his claim Tombstone. The site of many of the most dramatic and romanticized encounters in western American history and mythology, Tombstone also drew many far more temperate adventurers to town than the Earps or the cattle-rustling "cowboys" of O.K. Corral infamy.[3] Tombstone mines proved to be fantastically wealthy for a short time, but by 1881 there was groundwater seepage, and a decades-long battle to drain the mines commenced. Keeping the mines both dry and profitable required a major, and ultimately unsuccessful, engineering project. After being a magnet for prospectors and hangers-on, Tombstone then became a home for mining engineers, many of whom sought opportunities to stay in Arizona after their work at Tombstone petered out.[4]

At the time of the Tombstone strike, copper was in demand for the emerging telegraph industry. Telegraph wires were spreading rapidly across North America, stretching even to the most remote western outposts. In addition, Thomas Edison opened his first central power station in New York in 1882. Incandescent lights were made available to consumers for interior use in 1886, by which time prominent public establishments such as the New York Stock Exchange, the Academy of Music in Chicago, and La Scala Opera House in Milan were electrically illuminated.[5] Although electricity was not yet a common aspect of everyday life, the budding electric industry was hungry for copper wire, and miners paid attention.

As the production of copper increased, however—more than 166 million pounds were produced domestically in 1885, an almost three fold increase over 1880—the U.S. price of copper dropped from $0.20 per pound to $0.11 per pound. This price drop was orchestrated in part by Michigan copper producers seeking to strangle their western competition. Price fluctuation did manage to drive many smaller companies out of business and cleared the way for the consolidated copper mines that emerged in the 1890s and grew to epic proportions in the twentieth century.[6] Base metal mining was ultimately more important to western North America in terms of population growth and employment than precious metals mining, but through the 1880s there were still more gold and silver mines than copper or iron mines.[7]

In this mixed atmosphere of optimism and uncertainty, copper producers from established regions began to explore deposits in more western territories. In particular, Montana, Utah, and Arizona were known to hold large ore deposits. The inhospitable climate; distance from rail lines and major transportation hubs; and continuing Apache raids on U.S. and Mexican nationals meant Arizona was the least appealing of these sites for many investors. Despite such challenges, rumors began to spread about prospectors who found large copper deposits in southern Arizona. Two mining districts gained particular renown by the end of the 1870s: Clifton-Morenci, in eastern Arizona; and the Warren district, in the Mule Mountains near the new town of Bisbee, about twenty-five miles north of the Arizona-Sonora border.

The important mine in the Warren district was the Copper Queen. A couple of U.S. Army officers discovered copper in the Mule Mountains in 1875 or 1877 (depending on who told the story). Both subsequently died in an Apache raid. George Warren, a local gambler and drunk for whom the district was named, officially claimed the first mine and briefly held a 1/9 share in what became the Copper Queen. It quickly became apparent that the Copper Queen was a rich mine with potential to pay spectacular dividends. A combination of good planning and luck enabled some investors from nearby Silver City, New Mexico, and a pair of highly skilled miners, Ben and Louis Williams, to develop it fairly rapidly. By 1880, it was known as a solid working proposition and a "must see" for visitors to Arizona interested in mining.[8]

What those visitors saw, of course, depended not only on their knowledge of mining but also on their subject positions within the industry. Where did they come from? Who employed them, and to what purpose? Were they hoping to invest or simply to gather information? In the early 1880s, two highly qualified scientific men were sent to tour the Copper Queen. John Daniell, manager of a Michigan copper mine, was unimpressed. The "present price of ore," he concluded, "is not a strong incentive for commencing copper mining in Arizona."[9] By contrast, James Douglas was impressed enough to purchase the mine in 1886 for Phelps, Dodge, and Co., which at the time was a lucrative import-export business with headquarters in lower Manhattan. The Copper Queen was the company's first successful investment in copper mining. By the 1920s, Phelps Dodge had divested itself of its original trading business and held a controlling interest in the largest copper districts in Arizona. It was responsible for producing fully 19 percent of the copper in the continental United States.[10]

That Daniell turned up his nose at the Copper Queen while Douglas negotiated its purchase cannot be explained by arguing that Douglas understood more about mining or that Daniell was the less competent engineer or made a bad decision. Indeed, the two men saw many of the same strengths and flaws in the property. Their divergent interpretations of the mine, however, suggest that Douglas and Daniell may have been well aware that their mine inspections required a *performance* of objectivity and that the report each wrote intentionally balanced site assessments with the needs of their employers in a manner that might best be described as negotiated objectivity.

The Mine Report

Mine reports, conducted by mining engineers and technical experts on behalf of interested parties, were among the most significant documents produced by mining engineers during the nineteenth and early twentieth centuries. The author of a 1910 field guide for mining investors observed that the one item that a mine prospectus *had* to contain to be of value was the report of "a mining engineer of good standing... [since] a mining engineer is fully aware that his standing well depends on the truthfulness of his reports." Therefore, unlike a mine agent or promoter, the engineer had little inclination to lead an investor astray.[11] Mining engineers themselves were well aware of the significance of their reports for the industry. Engineer John Hays Hammond observed that it was

> on the strength of the recognized integrity, ability, and successful experience of the engineer that capital is invested in mining enterprises. The success or failure of such enterprise determines the career and future value of the engineer. He risks his reputation when he submits a report to his client.[12]

Mine reports were, on the surface, straightforward technical documents, communicating in words information that could perhaps have been more efficiently conveyed visually, if maps were easier to transport or produce. Reports addressed the size and location of ore veins and lodes; the direction of mine shafts; distances from road and rail transportation; and the availability of wood and water. A comprehensive mine report was also filled with computations concerning the cost of labor; the cost of buildings; the cost of transportation; and the market price of copper. It might contain a lengthy history of the mine site, which could be a dry recitation of dates or a romantic narrative of times past. It might even contain an enclosure of one or more reports written by other engineers. Mine reports and surveys could be formal or informal, short or long, but a mine report was never an "objective" analysis of a mine's value, although within the mining industry it was necessary to discuss reports as if they were. Rather, a mine report was a carefully constructed argument in which the engineer weighed the perceived financial value of the mine, as assessed by his own survey, against the needs of his audience. This careful balancing act is what positioned a professional expert as an arbiter of knowledge. In the mining industry, the report was the marker of a mine engineer's expertise, a document that no other industry figure could reproduce.[13]

Just as the format of a mine report varied, so too did the way in which a mining engineer conducted the survey. A mine inspection could be a multiweek undertaking on a massive scale, overseen by a chief engineer, and encompassing a staff of engineers and a large crew of laborers and surveyors.[14] Alternately, an inspection could be a relatively informal matter undertaken by a single engineer in a single day. When Louis Janin traveled through Sonora in the 1860s, he kept a notebook to record details of the mines he visited and drafted formal reports for investors. He collected samples and interviewed the mine owners or managers regarding ore deposits. Janin also took detailed notes on the precise workings of the mine, including observations about upcoming improvements. He noted, for instance, that "the ore is brought . . . by burros, but may be brought in a cart when road is completed." He also paid close attention to ore handling, describing the processing equipment and documenting the number of laborers and whether they were paid by wage or contract.[15]

Janin also made subjective observations. At the Mina Prieta, a silver mine near Cucurpe, Janin mused that his assays appeared to underestimate the value of the ore. Despite what he considered to be this flaw in the data he was collecting, he decided that he would still "follow my plan of taking ore along the vein in any given point."[16] This hastily written note underscores the essential conundrum of a mine inspection. Designed to present large amounts of data clearly to the reader, the validity of a mine report rested on the engineer's analysis of the size and value of the ore body. Mine shafts, outcroppings, and "ore in sight"—ore that was visible to the naked eye—were all measurable.[17] Questions about the extent of the ore body, such as how far it extended beyond the current shaft and which direction would be best for further exploration, could be answered with less certainty. Yet these were the central questions a mining engineer needed to answer in conducting his inspection. To solve this problem, mining engineers relied on two things: assays of ore samples and geological knowledge and experience with the physical presentation of ore.[18]

Sampling

The value of collecting samples for assays is fairly self-evident. The only way to know how to process ore to produce valuable metals was by knowing the composition of the ore. Samples were collected at regular intervals throughout a mine site, either by scraping ore from the walls of a mine or by drilling sample holes from the surface into a suspected or known ore body. Once collected, assays were sometimes performed on site at a mine, although the possibility of corrupting the sample was high. Often samples were packaged and sent to an assay office staffed by a mining or metallurgical engineer, and the mining engineer waited to write up his final reports until he heard from the assay office and could include the information with his findings.

While assay results were critical to mine reports, there was little standardization in the presentation of results. Most commonly, engineers simply stated the average from a given ore face or mine site. Withholding data points in this way gave the reporting engineer an advantage. Many mine investors were quite knowledgeable about mining, but they were not engineers and sought information from their technical experts that would allow an easy decision about a given prospect. If it was known that *x* amount of ore, assayed at *y* percentage copper, was needed for a mine to turn a profit, then an engineer could present his assay results in precisely that formulation. Similarly, engineers knew that a positive mine report would be widely circulated among investors and other mining engineers. Keeping the details of the results private could be to his benefit if an engineer wanted an ongoing association with the mine. A mine engineer who had performed his own survey and was thus familiar with the peculiarities of a given mine was in an excellent position to negotiate with the owners, regardless of who they were.

Assay samples presented another problem in that they were also the easiest place for a mine inspection to be falsified. Investors were aware of the inherent risks and hired mining engineers to mitigate them.[19] However, hiring a mining engineer could not completely secure an investor from fraud, in part because of the relative ease with which assay samples could be tampered with or "salted." This was particularly true in the precious metals mining that dominated the borderlands prior to the 1880s, although salting was not unknown in copper mining.[20]

The common form of salting was by adding value to samples, sprinkling ore with gold dust, for instance, or trading some of the local ore for samples from a different mine. More elaborate schemes could also be perpetrated, involving both tampering with a mine site and stock-market fraud. The most infamous of these was the so-called Diamond Hoax, rumored to have occurred in northern Arizona. Mining engineer Henry Janin was led, blindfolded, by a pair of prospectors and their chief investor. During a rushed and uncomfortable survey of the claim, Janin recorded finding several gemstones, occurring always with other gemstones known to be found in proximity to diamonds. Word of this remarkable find spurred a miniature diamond rush. A few short weeks after Janin's trip to the mine site, it was all unmasked as a fraud. The site itself had been salted, the diamonds planted for Janin to find.[21]

Ore Formation and Appearance

All knowledge of sampling practice, however, would be moot if the mining engineer did not also have an idea of the shape of the ore body. Mining engineers thoroughly studied the most up-to-date theories of ore formation in order to learn how to theorize how such bodies might look. The classic appearance of an underground mine is either a single or a shattered crevice containing ore, with a vein of metal-bearing ore extending through the natural cracks in a rock's formation. The ability to anticipate the depth, direction, and extent of these veins was one of the most important skills for which mining engineers were paid.

The other critical piece of information about an ore deposit was its value. Ore was generally classified in three categories. First-class ore is rare: these are ores that are found in large, dense bodies, usually as wide veins or in pockets, that are relatively simple to identify and extract. The valuable copper fields at Lake Superior exemplify first-class ore. Second-class ore is formed when minerals are dissolved in water under great pressure at great depth in the earth. Then, if the water solution makes its way to the surface (as part of a geyser, for instance) the ore remains behind in the cracks and fissures of the rock. The copper mines at Bisbee, for instance, were formed when metals were precipitated into the seams and fractures in the porous limestone. Third-class ores are deposited by water that percolates down from the surface. Because the ores are dispersed and low-grade they were considered essentially inextricable for most of the nineteenth century. Although any grade of ore could manifest as large veins, shattered veins, or cavernous deposits, the appearance of ore on the surface of the ground often gave mining engineers a hint of how the ores continued underground. Skilled engineers triangulated between surface appearance, sampling results, and knowledge of ore formation to predict the appearance of ore below the surface.[22]

The Copper Queen Mine

James Douglas, 1881

In 1880, William Dodge, of Phelps, Dodge and Company, hired James Douglas to visit Arizona to report on copper mining opportunities in the territory. Douglas remained there for the better part of a year, first examining the Eureka, a mine near Jerome.[23] Although the mine had a good appearance, "the size of the body, its inaccessible position, and its great distance from the railroad, were forbidding factors," and Douglas decided it was not a good investment.[24] Douglas then headed south to Morenci, Arizona, to report on the property of the Detroit Copper Mining Company. On Douglas's recommendation, Dodge did invest in this property. Although Phelps Dodge had existed for almost a century as a merchant and trading house, this was its first copper mine.[25] Finally, Douglas traveled still farther south to the Mule Mountains outside Bisbee to examine the copper mines of the Warren District.

In the Warren District, Douglas's original survey of the Copper Queen Mine, in 1881, did not begin auspiciously. The Copper Queen was not really in view as a possible purchase—he came to the district to view the Atlanta claim, which adjoined the Copper Queen. There appears to have been a miscommunication between Douglas and another mining engineer, Arthur Laing, about the kind of samples that Laing and his associates should collect from the Atlanta. Laing's assistant attempted to "supply the deficiency" of their survey "from his notes and field sketches."[26] These deficient notes, along with Douglas's physical inspection of the mine, informed Douglas's opinion of the claim. Douglas liked the look of the Atlanta, and although he cautioned that he had only taken a very superficial look at the property, Phelps Dodge purchased the claim in late 1881. Douglas took a one-tenth interest in lieu of cash payment.[27] The visit to the Copper Queen was almost an afterthought, undertaken on behalf of eminent Yale geologist Benjamin Silliman, who asked Douglas to send him a report on the mine since he was already in the area.

Douglas's report on the Copper Queen, in contrast to the apparent superficiality of his report on the Atlanta, was extremely thorough. He began by detailing the appearance of the mine, at which the bedrock limestone was "traversed... by masses of quartzite... [in which] many small bunches of copper have been found *but none have proved permanent*" (my emphasis).

More encouraging was Douglas's estimate of the ore-in-sight. Even here, Douglas qualified his observations, claiming that although his calculations were not necessarily accurate he had great confidence in them. Indeed, a reader would have great confidence in his calculations, which are impressively comprehensive. He described in excruciating detail how he came to his conclusions, painting a clear picture of the size of the mine; the appearance of the ore on each face of the cave; the appearance of the ore on each side of the drift; the height of the cave as it could be seen; and the depth of the ore as it might reasonably be concluded from the exploratory drifts that had already been cut. In conclusion, he estimated the value of the ore-in-sight at $971,660—an unequivocal fortune.[28] However, as noted in a letter to Silliman, Douglas thought the cost of labor was "unreasonably high," and the true extent of the deposit was difficult to assess with any accuracy.[29]

At the end of 1881, Phelps Dodge purchased the Atlanta claim, but negotiations to purchase the Copper Queen failed when the owners thought they could get more money than Phelps Dodge and Douglas were offering.[30] The Copper Queen's proximity to the Atlanta unnerved Douglas. He was concerned that Phelps Dodge's work on the Atlanta, which proved unprofitable in the early 1880s, would run up against the ore body of the Copper Queen, causing Phelps Dodge to lose its stake due to the so-called "apex rule." A piece of legislation peculiar to U.S. mining law, the law of the apex permits a miner who holds the outcropping of a given vein to follow that vein extralaterally into a neighboring claim.[31] The apex law, initially a local custom originating with the Comstock Lode, was brought into federal law with the Mining Law of 1872 and was almost universally loathed by mining engineers, who correctly blamed it for creating lawsuits and distractions from the real business of mining.[32] Mining engineer Eben Olcott neatly summarized the difficulty for Phelps Dodge of owning the Atlanta but not the Copper Queen.

> In regard to the Copper Queen ore dipping into your property, if it does and they [the CQ interests] own the out crop of such vein or deposit they can claim all that lies in your property by U.S. Law.[33]

Douglas's fears were indeed justified, as the Atlanta and Copper Queen did enter into litigation over lateral rights. These varied circumstances meant that over the next four or five years Douglas kept an eye on the Copper Queen, making several formal and semiformal inspections. He continued to believe that the ore was of generally high quality, and his continued good relations with the on-site managers, Ben and Lewis Williams, meant that Douglas remained informed of the state of the Copper Queen.

John Daniell, 1883

Two years after James Douglas's initial trip to Arizona, John Daniell also visited the territory at the behest of J. W. Clark and A. S. Bigelow. Clark and Bigelow were Boston-based investment partners in the Tamarack and Osceola Mining Companies, two copper mines on the shores of Lake Superior where Daniell had worked for several years. The Osceola abutted the Calumet and Hecla, the enormous mine that set the standard for copper production in Michigan. The Tamarack was slightly south of those mines on the Keweenaw peninsula. Under Daniell's management, the Osceola, a relatively poor neighbor of the Calumet and Hecla, mining in part the same lode, made its first profits after years of desultory production. In later years, Daniell was also known as a significant innovator of excavation techniques. Shortly after his return from the West, Daniell pioneered a new system of "deep-level" mining, sinking a shaft down to the unheard-of depth of 2,300 feet.[34] In 1883, however, that achievement was still in the future, and Clark and Bigelow, feeling the pinch of the western mines on the profits of Lake Superior copper production, sent their top technical expert west on a fact-finding mission. Daniell toured the copper mines near Butte, Montana, and then made a brief trip to Arizona. Butte so impressed him that he spent longer than intended in Montana and was pressed for time when he reached Arizona, such that he chose only to visit the most well-established copper mine in the territory—the Copper Queen.[35]

Like James Douglas, Daniell first described the location and outward appearance of the mine in his report, drawing attention to the shape of the formation, the appearance of several quartzite outcroppings in the limestone bedrock and the main deposit's angle of descent from the surface (forty-five degrees). But Daniell was frustrated with his inability to fully describe the mine. He complained, "It is not easy to convey one's ideas of this mine in writing." The easier way to describe it to Clark and Bigelow, who had broad experience in mining, was by comparing it to mines that they knew. The mine's opening, Daniell explained, resembled that of an iron mine in the Marquette district in Michigan. The limestone bedrock, on the other hand, was similar to the "mineralized lodes of our Lake Mines." The ore composition, too, was similar in copper content to the Quincy Mine Lode, another mine on Michigan's Keweenaw Peninsula with which Clark and Bigelow were certainly familiar. The appearance of the ore, as a cavernous deposit within the limestone, resembled the appearance of the interconnected ore chambers at the Eureka Mine in Nevada. Unlike the Eureka, however, the Copper Queen appeared to only consist of the single chamber.[36]

Daniell further described the workings at the mine in some detail. At the time of his visit, the Copper Queen had tunnels at four levels. The fourth level he dismissed altogether, as "although a cross cut had been driven South of [the] shaft some distance, nothing but limestone is exposed." The same quantity of ore—nothing—was apparent in a shaft branching east, "although in the level over [level three] ore is continuous for that length." Furthermore, the bottom of the mine was filled with water. Daniell thought it unlikely that any exploratory diggings would go below the fourth level, since it appeared to him that the third level, which he observed to be lean of copper and heavy in iron, was the bottom of the deposit.[37] He quoted the superintendent (likely Ben Williams, an experienced miner who was not an engineer and who worked at the Copper Queen from the 1870s until the early twentieth century) as saying that there was one and a half years of ore left at the current rate of extraction.

Although Daniell's observations about the appearance of the mine itself were somewhat elliptical, his calculations of the cost of running the mine, based on the information he was able to obtain from the superintendent, were quite specific. He reported the ore production of each level in the previous year; the value of that ore at point of sale; the cost of refining a ton of ore; and the purity of the ore, which averaged around 12.73 percent the previous year, although at the time of his visit the yield was averaging only 10 percent. Square-set timbering supported the mine shafts. To support this production, the Copper Queen was going through wood at the astonishing rate of 80,000 feet per month at a cost of $1 per ton. Wood was one of two significant costs for the mine. The other was fuel for running the furnace. Coke, brought from Trinidad, New Mexico, cost the Copper Queen $22.50 per ton.[38]

Daniell was very impressed by the fact that the Copper Queen furnace was on site. Likewise, he also appreciated the fact that the ore body was large, easily worked, and relatively rich. The ore separated easily, and he considered that the slag, or waste rock, was quite poor—a good sign, as it meant that the refining process was efficient. He noted that the "facilities for marketing" the copper, meaning arrangements to transport the copper ore out of Arizona, were unequalled in Arizona or elsewhere in the West. The Copper Queen was only forty miles from the nearest railroad—a relatively short distance.[39] Yet his overall analysis was bleak. "Rapidly nearing the point where they will find no money in copper," Daniell observed. The Copper Queen was not a good investment. He concluded his report by declaring that the "present price of ingot is not a strong incentive for commencing copper mining in Arizona."[40]

Phelps Dodge, 1886

By 1885, the owners of Phelps Dodge had made a significant investment in the Atlanta property to little effect. In spite of this lack of success, and because of their ongoing investment there, William Dodge and D. Willis James, the managing partners of Phelps Dodge, remained interested in purchasing the Copper Queen, and James Douglas made sure that he remained informed of the workings, product, and prospects of the mine.

Despite John Daniell's assertion in 1883 that the Copper Queen only had one-and-a-half-years' production left in it, Douglas was able to report to Phelps Dodge in New York that the copper production in May 1885 was the largest production of copper from the Copper Queen furnaces in the history of the mine. At this time, the Copper Queen was producing an average of forty tons of ore per day, of 8 percent copper, a significant decrease in quality from the 12–20 percent ore Douglas saw on his initial visit in 1881.[41] However, Copper Queen ore still appeared near exhaustion, and when superintendent Ben Williams struck an extension of the Atlanta ore vein on Copper Queen property, the owners of the two claims finally decided to join their operations. The Copper Queen Consolidated Mining Company was established in 1885.[42] By 1887, Douglas had instituted efficiencies of scale and persuaded Phelps Dodge to build a railroad, placing the Copper Queen on the path toward profitability and market dominance.[43]

Rhetoric

The Rhetoric of Mine Reports

Because of the significance of mine reports in the professional lives of mining engineers, these reports tended to be extremely conservative documents, despite the significant investments of capital that they frequently endorsed. Engineer Louis Janin explained this conundrum, noting that mining engineers "must be cautious in conclusions but bold in executing; an expert is cold nosed."[44] In a similar vein, another mining engineer explained that a consulting engineer had to be "as cold-blooded and as unenthusiastic as a clam."[45] For a mining engineer, writing a mine report was an exercise in managed risk. Mining was a speculative industry, and the principle reason engineers were hired was to increase the odds that an investment would succeed. The highest praise a mining engineer could receive was that he was cautious and conservative. Mining engineers' status as educated professional men rested upon their deployment of technical expertise, the most obvious manifestation of which, in the late nineteenth century, was the geological and mechanical knowledge that they obtained at university.

Obvious risk-taking suggested that such hard-won expertise was worthless.[46] Yet the nature of the industry meant there was always the possibility that despite his best efforts a consulting engineer would make a mistake. After all, engineers understood that it was

> on the strength of the recognized integrity, ability, and successful experience of the engineer that capital is invested in mining enterprises. The success or failure of such enterprise determines the career and future value of the engineer. He risks his reputation when he submits a report to his client.[47]

Therefore, mining engineers themselves had a rhetorical challenge when constructing their reports. Sounding authoritative while acknowledging that expertise had limits was a difficult balancing act.

Mining engineers made frequent use of such terms as "cautious," "conservative," and "careful" in their reports, even when the conclusions thus qualified were clearly absurd, as when an engineer concluded that the ore reserves at the Mulatos Mine in Sonora were "by careful calculation, simply inexhaustible."[48] More often, caution was asserted so as to highlight weaknesses in the engineer's own work. In a report on a mining property in Yuma County, Arizona, for instance, mining engineer John Church reiterated his point so that it would be perfectly clear: "No careful survey of these claims has ever been made.... A careful survey is a necessity before any purchase can be made." Church had inspected the mine in question but did not conduct a full survey. But in the cover letter accompanying his highly qualified report, he demonstrated that despite all the qualifications he thought this mine was worth taking a risk. Although he noted that there was reported to be $450,000 net profit to be made from the ore-in-sight, he suggested that it might be worth paying as much as $800,000 "b/c what's not in sight may be equally valuable."[49]

Church's equivocation about the ore-in-sight—the only calculation a mining engineer could make with any great authority—underscores the inherently qualitative nature of mine reports. Once again, comparison of the initial Copper Queen mine reports of Douglas and Daniell illustrates how mining engineers presented information in distinct ways depending on how they wanted their reports to be received. Douglas was meticulous in presentation and made no statements that could be interpreted as equivocation. Of the point where the tunnel from the surface met the ore-filled cavern that was currently being mined, he stated that "to the west... [the tunnel] is throughout in good ore," and furthermore, that the eastward drift "has cut through the ore body." The samples he takes from these drifts, then, yielding 20 percent copper and 15 percent copper, are understood to be from the main ore body and further understood to be rich. Douglas's interest in the neighboring Atlanta mine as well indicates his belief that the Copper Queen ore body might extend laterally.

Daniell, on the other hand, writing of the same point in the mine, where the main tunnel met the primary ore body, stated, "It may be unnecessary to say that near the surface this is quite rich for copper carrying fully 20%." He then moved on to complain, "It is not easy to convey one's ideas of this mine in writing," and proceeded to list the mines to which he would compare the Copper Queen. Daniell thus easily conveyed that he found the mine to be peculiar, and, by extension, untrustworthy. This opinion was reinforced by his conclusion that "no mine [in Arizona] can be counted on for a foot in depth or length beyond where ore can be seen...." Lest it seem that the differences between Douglas's and Daniell's views of the mine are solely the differences between 1881 and 1885, note that when Douglas visited the Copper Queen in 1885, a few months prior to the Phelps Dodge acquisition, he reported that a large amount of ore was found in discovery—i.e., tunnels dug beyond the ore-in-sight—and that in the previous month the furnace made "the largest run...net of copper—on their record."[50] In other words, Douglas reported that the exact situation Daniell described as impossible—finding ore beyond what was in sight and proving that the Copper Queen had longevity—was in fact possible.

Perspective

James Douglas and John Daniell visited the same mine with the same goal—determining whether it merited further investment on the part of their employers. Yet they wrote vastly different reports. Douglas concluded that the mine was worth considering and attempting to acquire; Daniell concluded that although not without interest, the Copper Queen was a dubious proposition and one that was best to avoid. That two technical experts, both highly skilled engineers with significant experience surveying mines and operating mines, could come to such vastly different conclusions underscores how mining engineers manipulated their reports while simultaneously pointing to them as evidence of their own objective expertise.

Some of the difference in the Douglas and Daniell reports can clearly be traced to their divergent perspectives on Arizona mining in general. In 1883, John Daniell was probably one of the most experienced copper miners in North America. As a veteran manager of copper mines in Michigan, he was accustomed to thinking of mines with low-grade ore as failing. By the 1880s, the ore from the Osceola, for instance, averaged around 1.2 percent yield, as opposed to the double-digit averages from the ore body at the Copper Queen.[51] In Michigan, such low yields usually indicated that an ore body was playing out. Daniell's morose assessment of the likelihood that the Copper Queen contained more ore than was "in sight" in 1884 would naturally lead him to think of the Copper Queen as a disappointing prospect. The mines that he worked in Michigan, the Osceola and the Tamarack, were significant mines. The Osceola, for instance, produced 2,089 tons of copper with a market value of about $762,846 in 1881 alone.[52] This production rate was only about 25 percent higher than the total ore in sight at the Copper Queen that same year, according to Douglas's calculations. Thus, by the standards of the upper Midwest, the Copper Queen could not be said to be either productive or accessible.

In addition to his work experience with Lake Superior copper, Daniell's more immediate experiences touring mines in Butte likely also affected his generally dismissive view of the Copper Queen. In 1884, Butte was shipping an astronomical amount of copper ore—by some estimates, over $10 million annually.[53] Even the comprehensive *Mineral Resources of the United States*, issued by the USGS in 1885, was cautious about the continued productivity of Butte, stating that "the heavy shipments of rich ores may be expected to decline very rapidly" and that the future value of the district would be in silver.[54] By contrast, Daniell was rhapsodic about the mines he viewed in Montana, describing the Anaconda as "undoubtedly the Copper Mine of the West standing as far in advance of any other mine we saw as the Cal + Hecla does ahead of the other Lake Superior Mines."[55] The Calumet and Hecla routinely produced six or seven times more copper each year than neighboring mines. On Daniell's recommendation, Clark and Bigelow put together a syndicate that became the Boston and Montana and purchased several mines in the vicinity of Butte, becoming the second largest producer of copper in North America.[56] Although this mining group later took shares in a handful of Arizona mines, the bulk of its work remained in Michigan and Montana, and it was never a serious operator in Arizona or the borderlands.

James Douglas had a completely different frame of reference for considering the opportunities presented by the Copper Queen. For instance, Daniell's experiences in Michigan had him working one of the most spectacular copper lodes for investors who were familiar with copper mining and had significant resources to support a mine and its operations. Douglas's most recent mining experience ended in the failure of a smelting enterprise in Pennsylvania, and he was eking out a living as an independent mine

This meeting of the Masonic Grand Lodge of Arizona, in a cavern at the Copper Queen Mine in 1897, demonstrates the scale of the excavation undertaken by Phelps Dodge. *The Masonic Grand Lodge of Arizona meeting in the cave of the mine of the Copper Queen Consolidated Mining Co. at Bisbee, Arizona, November 12, 1897.* A. Miller, Library of Congress, https://www.loc.gov/item/99472628/, Nov. 7, 2017.

inspector and surveyor. These job trajectories provided divergent ideas about what constituted a "paying" mine. Perhaps James Douglas was simply more desperate than John Daniell, and so he was prepared to see opportunity where none was clearly presented. The author of one book about the history of copper mining in the United States asks, "Did James Douglas make the Copper Queen or did the Copper Queen make James Douglas?"[57] The implication is clearly that Douglas's initial success was simply luck. This conclusion, while it makes for a fine narrative, is analytically unsatisfying, as it suggests that Douglas did not understand the possibilities inherent to the Warren district mines. Douglas's Copper Queen reports, however, show that he clearly did. Moreover, he repeatedly stated that the Copper Queen interests ought to purchase the Atlanta, and he urged Dodge and James to consider purchasing the Copper Queen, demonstrating his belief that the two claims ought to be worked in concert.

Another significant distinction between the two men was that while Daniell was a stranger to Arizona and only had time to investigate the Copper Queen, Douglas had already been in the territory for close to a year before he visited the mines at Bisbee. The advantages of the Copper Queen and its neighboring claim, the Atlanta, noted by both Douglas and Daniell—relative proximity to a railroad; significant supply of wood; a readily accessible ore body; and decent on-site refining works—stood in contrast to what was available in most of the rest of Arizona. For instance, prior to viewing the Copper Queen, Douglas visited the mines of the Detroit Copper Company, outside Clifton, Arizona. Douglas liked these mines well enough to suggest William Dodge take a share in them, although he did not think it reasonable at the time to attempt to purchase and run them. Of significant concern at Clifton was the lack of available timber. Ore had to be left unexploited in order to support the roofs of the shafts and stopes, which significantly depleted possible earnings.[58] By comparison, the available raw materials at the Copper Queen, which Daniell thought were not rich enough to justify investment, likely appeared abundant to Douglas, who had a broader basis of comparison for the resources of Arizona.

Finally, Douglas and Daniell wrote for two very distinct audiences, and this, possibly more than their different levels of knowledge of Arizona metals, or their dissimilar experiences working copper mines, determined the outcome and tone of their reports. Daniell's employers were already successful copper investors, well on their way to becoming western mining magnates. He had to judge whether the resources at hand would be better used in Arizona or to enter the fray in Butte and attempt to carve off a piece of the Anaconda wealth. Douglas, on the other hand, worked for men who knew a great deal about copper trading but were just starting out in mining. The relative logistical challenges of mining in Arizona may have seemed less overwhelming to Phelps Dodge, a company with no experience in the relatively easy mining in Michigan.

To Daniell, whose own professional life was shaped in the Lake Superior copper mines, the work in Arizona presented an entirely new set of professional challenges. Given his experience with Lake Superior mining, it is apparent why the Bisbee mines appeared an untenably weak proposition. In writing their respective reports, Douglas and Daniell each enacted a performance of objectivity that served to guide their employers' investment decisions while satisfying their own professional responsibility—and the needs of the technocratic bureaucracies in which they operated—to provide accurate analysis of a mining property.

Notes

1. [Sonora Exploring and Mining Co.], *Sonora...1856*; [Sonora Exploring and Mining Company], *Third Annual*; "Silver and Copper Mining in Arizona," *Mining Magazine*, 2nd series 1, no. 1 (November 1859); [Santa Rita Silver Mining Company], *Second Annual*; [Louis Janin Jr.], *Reports*. The concerns voiced in these reports carried over into more official publications of the 1870s. See Rossiter Raymond, *Silver and Gold: An Account of the Metallurgical Industry of the United States, with Reference Chiefly to the Precious Metals* (New York: J. B. Ford and Company, 1873), 331–340.

2. Robert L. Spude, "A Land of Sunshine and Silver: Silver Mining in Central Arizona 1871–1885," *Journal of Arizona History* 16, no. 1 (1975):31–32, 58.

3. Rickard, *A History of Mining in America* (New York: McGraw-Hill, 1932), 275; Truett, *Fugitive Landscapes*, 61–63; St. John, *Line in the Sand*, 70.

4. Hyde, *Copper for America*, 112; I discuss the later engineering history of Tombstone in greater detail in chapter 6.

5. David Nye, *Electrifying America: Social Meanings of a New Technology* (Cambridge, MA: MIT Press, 1990), 31–32.

6. Hyde, *Copper for America*, 59–60, 81; Albert Williams Jr., *Mineral Resources of the United States, 1883–1884* (Washington, D.C.: Government Printing Office, 1885), 326–327. Hyde calculates 60 million pounds of domestic copper produced in 1880, and 166 million in 1885. By 1885, only about half the copper mined in the United States came from Lake Superior; the rest came from Montana, Arizona, Nevada, and points west.

7. Hyde, *Copper for America*, 80.

8. On the early history of the Copper Queen, see John T. Howard, "Army Officers Located First Copper Claims on Mules...," Copper Queen folder, box 1, MS Cleland-Phelps Dodge, The Huntington Library, San Marino, CA (hereafter C-PD); Robert Glass Cleland, *A History of Phelps Dodge* (New York: Knopf, 1952), 84–87; Rickard, *A History of Mining in America*, 282–284; Truett, *Fugitive Landscapes*, 67–68; Hyde, *Copper for America*, 123.

9. John Daniell to J. W. Clark, December 3, [1883], letterbook 15, John Daniell Papers, The Huntington Library, San Marino, CA (hereafter Daniell Papers).

10. Hyde, *Copper for America*, 162–163.

11. Etienne Ritter, *From Prospect to Mine* (Denver: Mining Science Publishing Company, 1910), 73.

12. Hammond, *Autobiography*, 148.

13. Ash, *Power*, 15.

14. A description of a mine survey in Argentina can be found in Charles Hoffmann to F. Perugia, Esq., 1899, folder 28, box 2, MS 3163, Ross Hoffmann Collection, American Heritage Center, Laramie, WY (hereafter RH). There are numerous examples in the archival sources of the details mining engineers looked for when making reports; see Arthur Laing to James Douglas, July 18, 1881, folder 29, box 3, JD; James D. Hague to Santa Eulalia Mining Co., May 20, 1885, L-9, JDH. On contractual agreements to examine mines, see [Agreement between James D. Hague and Perkins, Livingston, and Co.], July 8, 1879, and [Agreement between James D. Hague and Phelps, Dodge, and Co.], K-5; Ellsworth Daggett to James D. Hague, August 3, 1884, M-13, both JDH; [unknown] to George J. Hoffmann, March 13, 1900, folder 29, box 2, RH.

15. [unpaginated entry, ca. spring 1864], Louis Janin Diary, Sonora, 1863, HM 64294. An example of Janin's formal reports can be found in [Louis Janin Jr.], *Reports*.

16. [unpaginated entry, ca. spring 1864], Louis Janin Diary, Sonora, 1863, HM 64294.

17. Nystrom, *Seeing Underground*, 25.

18. In *Seeing Underground*, Eric Nystrom argues that mining engineers understood mapping and this knowledge differentiated them from knowledgeable working miners. He makes a strong case that this was true after about 1910, but in the late nineteenth century mine valuation was harder to quantify, as many of the mapping tools that were available to twentieth-century mining engineers had not yet been devised.

19. Balch, *Mines*, 859–864.

20. Sarah E. M. Grossman, "Mining Engineers and Fraud: The U.S.–Mexico Borderlands, 1860–1910," *Technology and Culture* 55, no. 4 (2014):821–849.

21. There was an enormous amount written about the Diamond Hoax in the 1870s, and historians and journalists have remained interested in the event. For a taste of the contemporary coverage, start with "Diamond Fields on the Pacific Coast," *Mining and Scientific Press* 25, no. 5 (August 3,1872); "MUST the New Diamond Fields," [*sic*] San Francisco *Sentinel* (1872); "The Arizona Diamonds," *Chicago Tribune*, (August 9, 1872); "The Diamond Excitement, in All Probability, a Great Swindle," *Chicago Tribune* (August 19, 1872); Clarence King, "Report of Clarence King, United States Geologist, to the Board of Directors of the San Francisco and New York Mining and Commercial Company," *Engineering and Mining Journal* 14 (December 10, 1872):377–380 [which was also published in both the *Chicago Tribune* (December 5, 1872) and the *New York Times* (December 5, 1872)]. Later and more measured accounts have been provided by, among others, Hubert Howe Bancroft, *History of Arizona and New Mexico, 1530–1888* (San Francisco: A. L. Bancroft, 1889), 1591–1592; Jeremy Mouat, "The New Mexico Origins of the Diamond Hoax," paper presented at the Silver City Museum, Silver City, NM, June 9, 2010; A. J. Liebling, "The American Golconda," *New Yorker* (November 16, 1940):49–64; Martha Sandweiss, *Passing Strange: A Gilded Age Tale of Love and Deception Across the Color Line* (New York: Penguin, 2009), 68.

22. Ritter, *From Prospect to Mine*, 21–36; Richard Francaviglia, "Copper Mining and Landscape Evolution: A Century of Change in the Warren Mining District, Arizona," *Journal of Arizona History* 23, no. 3 (1982):270. That metal ores were somewhat predictable below-surface made metals mining at the close of the nineteenth century different from the other kind of prospecting that was gaining value: searching for oil. Geologists and oil men attempted to take the lessons of generations of miners to heart in seeking the "black gold," but they were perpetually thwarted by the fundamentally different nature of oil from both metals and other fossil fuels such as coal. Brian Frehner, *Finding Oil: The Nature of Petroleum Geology, 1859–1920* (Lincoln: University of Nebraska Press, 2011), 82–84.

23. Cleland, *A History of Phelps Dodge*, 95–98.

24. James Douglas, "Autobiographical Recollections, III," p. 1, folder 3, box 2, LD; [Biographical narrative of James Douglas], 150–151, folder 2, box 1, JD.

25. Carlos A. Schwantes, *Vision and Enterprise: Exploring the History of Phelps Dodge Corporation* (Tucson: University of Arizona, 2000), 57–59; [Lewis Douglas],

[Biographical narrative], 175–179, folder 3, box 1, JD. On the finances of Phelps Dodge between 1880 and 1950, see Cleland, *History of Phelps Dodge*, appendices 11–14.

26. Arthur Laing to James Douglas, July 18, 1881, folder 29, box 3, JD.

27. Cleland, *A History of Phelps Dodge*, 98–99. The apparently "superficial" report on the Atlanta that Cleland described is either no longer extant or not available to researchers. Hyde, *Copper for America*, 125.

28. James Douglas Report on Copper Queen, 1881, transcribed by Robert Glass Cleland, folder 1, box 1, C-PD.

29. James Douglas to Benjamin Silliman, February 8, 1881, folder 86, box 5, JD.

30. Truett, *Fugitive Landscapes*, 69; Edward Reilly to James Douglas, February 14, 1881, folder 29, box 3, JD.

31. Gordon Bakken, *The Mining Law of 1872: Past, Politics, and Prospects* (Albuquerque: University of New Mexico Press, 2008), 5–6, 61–71.

32. Rossiter Raymond et. al.

33. Eben E. Olcott to William E. Dodge and D. Willis James, September 20, 1881, folder 30, box 3, JD.

34. T. A. Rickard, *The Copper Mines of Lake Superior*, 17, 50, 156.

35. John Daniell to J. W. Clark, December 3, [1883], letterbook 15, Daniell Papers.

36. Information on the composition of the Quincy Lode can be found in Rickard, *The Copper Mines of Lake Superior*, 26–27.

37. John Daniell to J. W. Clark, December 3, [1883], letterbook 15, Daniell Papers.

38. Sam Truett describes the astonishing impact of the Copper Queen on the timber, water, and fuel resources surrounding Bisbee in *Fugitive Landscapes*, 71–77.

39. Daniell noted the good location of the Copper Queen mine on several occasions after his initial visit, most notably in John Daniell to J. H. Chandler, September 22, 1885, letterbook 15, Daniell Papers.

40. John Daniell to J. W. Clark, December 3, [1883], letterbook 15, Daniell Papers. Daniell revised his opinion on Arizona mines in 1890, after he undertook a comprehensive survey of the mines of the Arizona Copper Company in Clifton. Calling the report "about the hardest piece of work I ever tackled," he nonetheless reported favorably on the ore quality and prospects of the mine, then under the auspices of James Colquhoun and discussed in chapter 5. John Daniell to the Lewiston brothers, November 20, 1890, letterbook 24, Daniell Papers.

41. James Douglas to William E. Dodge, June 7, 1885, folder: Copper Queen Mine, box 1, C-PD.

42. Hyde, *Copper for America*, 126; Ira Joralemon, *Romantic Copper: Its Lure and Lore* (New York: Appleton-Century, 1936), 123–124; James Douglas, "The Copper Queen Mine, Arizona," *Transactions of the American Institute of Mining Engineers*, vol. 29 (1899):515.

43. Truett, *Fugitive Landscapes*, 71–72.

44. Louis Janin to J. Blythe, October 19, 1887, L2, Louis Janin Papers.

45. Cited in Spence, *Mining Engineers*.

46. Gunther Peck, "Manly Gambles: The Politics of Risk on the Comstock Lode, 1860–1880," *Journal of Social History* 26, no. (1993):701, 704.

47. Hammond, *Autobiography*, 148.

48. E. A. Brandon, "Mulatos Mineral Zone and Land Grant, Mexico," n.d., P-89, JDH.

49. John A. Church, "Condensed Copy: Report on la Fortuna Mine Yuma County, Arizona," P-70, JDH; John A. Church to H. L. Higginson, December 19, 1896, P-70, JDH.
50. James Douglas to William E. Dodge, June 7, 1885, Copper Queen folder, box 1, C-PD.
51. Williams, *Mineral Resources of the United States*, 331.
52. Balch, *Mines*, 1031.
53. Lingenfelter, *Bonanzas and Borrascas*, vol. 2, *Copper Kings and Stock Frenzies, 1885-1918* (Norman, OK: Arthur H. Clarke, 2012), 19.
54. The assumption that Montana copper production would soon falter was completely mistaken. Montana continued to produce over 40 percent of U.S. copper until the 1910s, when the vast porphyry deposits in Arizona began to be exploited in earnest. As discussed in chapter 6, by 1920 the productivity of these deposits outpaced Montana by a factor of three. Hyde, *Copper for America*, 81; Williams, *Mineral Resources of the United States*, 331.
55. John Daniell to J. W. Clark, December 3, [1883], letterbook 15, Daniell Papers.
56. Hyde, *Copper for America*, 85–86.
57. Joralemon, *Romantic Copper*, 114.
58. James Douglas notes on Frank Church properties, May 1881, folder 92, box 5, JD.

CHAPTER FIVE

CORPORATE CAPITALISM

Engineers and the Birth of Mass Mining

In 1913, an editorial in the *Engineering and Mining Journal* bemoaned the so-called "new miner" in the western mines, a man who was no longer a "jack of all trades and master of several." Rather, the editorial lamented, this new miner lacked skill, industry, and intelligence.[1] Historians Logan Hovis and Jeremy Mouat argue that such perceived decline in the quality of the mining labor force was caused by new engineering practices, particularly the spread of so-called "mass mining."[2] The working lives of mining engineers also changed with the adoption of mass mining. While this new system simplified the work of average miners such that companies could now hire unskilled labor, the work of mining engineers became in some respects more complicated. Technological change grew out of changes in the sociotechnical system of mining itself. The advent of electricity fundamentally altered the economics of copper mining. As demand for copper rose—slowly through the 1890s and with increased urgency after 1900—it became profitable to mine the massive, yet relatively low-grade copper deposits of the U.S.-Mexico borderlands.[3]

The interaction between the social world and technological change is at the heart of many scholarly debates surrounding the agency of technology. Introduction of mass-mining techniques at the turn of the twentieth century highlights how the mining industry functioned systemically, as changes within society seemed to lead inexorably to technological change.[4] Mining engineers developed mass mining techniques in answer to the new demand for copper, and in turn, the new technology had a profound effect on their work. The new processes changed the daily experience of engineers as well as the general work-flow of the industry. But the life experiences of mining engineers also altered, as the profession became less mobile—both in terms of job-to-job movement and in professional advancement.

Much has been written about the effect of mass mining on the average worker, generally focusing on the deskilling of the workforce. Where selective mining required skilled miners and mining foremen to operate pick-axes, explosives, and hand-drills to burrow tunnels along underground ore veins, nonselective mining was just that: nonselective. Huge quantities of dirt, rock, and low-grade ore were removed from the ground all together. Once extracted, small quantities of ore were separated from massive quantities of waste. In selective mining, an engineer might design a dig, but he generally left the quotidian operation to the foreman and skilled workers, while providing technical oversight or adjudicating important decisions during the progress of the dig. With mass mining, the mining engineer or engineers designed the entire dig, and the miners removed dirt, rock, and ore. There was little need for individual expertise or knowledge of ores and their properties on the part of miners or foremen.

In Arizona, which in 1910 produced more copper than any state in the nation, and throughout the U.S.–Mexico border region more generally, the shift to mass mining meant operations increased dramatically in size and number. By the early twentieth century, the regional industry had become larger, more industrialized, and more integrated into the economic life of the nation than seemed possible just a few short decades earlier.

An influx of immigrant workers from southern and eastern Europe and an increased number of Mexican workers flooded the region as low-wage laborers. Along with the new workforce population, the pressures of industrial organization also increased. Radicalized workers steadily gained a foothold in an industry already familiar with the complications of labor actions, although there was an extant division between the American Federation of Labor and the Western Federation of Miners. In addition, miners were distanced from one another by language, ethnicity, and nationality.[5]

The shift to nonselective mining in the western metals industry was orchestrated by a small cohort of mining engineers. Faced with rising demand for copper in the marketplace and declining purity in subsurface deposits, mining engineers were tasked with finding new ways of extracting copper from the quartz in which it was embedded throughout the borderlands. Far from being simply a new technological approach to mining, the implementation of mass mining restructured the mining industry. As a more technology-intensive than labor-intensive way of mining, it favored large, capital-rich companies over smaller organizations and accelerated consolidation of the mining industry.

The image that comes to mind when we think of mass mining is the open pit, but the term actually encompasses a broad approach to metals extraction that pushed the industry toward greater mechanization. Environmental historian Tim LeCain has aptly named nonselective mining "mass destruction," as it constituted an inversion and adaptation of the mechanized production implemented by factory managers elsewhere in the United States. In mining, the work being standardized was the dismantling of the landscape, as opposed, for instance, to the construction of a Model T. LeCain's larger point is that techniques of mass destruction so heavily mediated the relationship between mine workers and the environment that the conditions of labor became much more dangerous, as technological solutions were sought for the environmental problems caused by large-scale earth-moving ventures.[6]

On the surface, such a full-scale embrace of technology might seem to benefit the workplace status of mining engineers, raising the value of their technical and technological expertise ever higher in the minds of capitalists and investors. But for individual engineers, the reverse was often true. Rather than gaining power and leverage within mining corporations, a growing number of engineers actually worked within engineering units, which permitted much less professional advancement. Mass mining also reduced opportunities for engineers to travel extensively, a major feature of mine-engineering identity in the nineteenth century that remained a critical aspect of the romantic vision of the profession and drew young men well into the twentieth century.

The story of the rise of mass mining often begins the way LeCain begins his narrative of the environmental history of Butte, Montana, with mine engineer Daniel Jackling's design of the open-pit mine at the Guggenheim-financed Bingham Canyon Mine near Salt Lake City in 1908. While the dramatic open cuts at Bingham Canyon did indeed presage a more aggressive era in copper mining, Jackling was not the only mining engineer experimenting with the new techniques. Rather, he was the most successful of a cohort of mining engineers who were struggling to solve the problems posed by the low-grade western porphyry coppers—not simply how to remove ore from the ground but how to profitably process commercial copper once the ore was on the surface.[7]

Early Mass Mining

In the borderlands region, as in the western United States more generally, mass mining was adopted in response to specific problems posed by the copper ore there. One characteristic of copper is that it bonds with numerous other elements, and thus copper ores are rarely found in a pure state. In the borderlands region in the 1890s, pure native copper ore was present in quantities that were profitable for commercial production only at Santa Rita, New Mexico.[8] Far more typical of the rich mines in the southwest was the Longfellow Mine in Clifton, Arizona, which held veins of chalcocite, a copper ore with a relatively high copper content—generally between 55 and 75 percent—embedded within limestone. Still more mining districts in the Southwest presented completely or primarily within porphyry, or igneous, rock. Sometimes veins of chalcocite or another relatively rich ore were embedded within the porphyry. For example, the Detroit Mine had a porphyry vein that was over a hundred feet wide at the surface. Usually, such relatively rich surface veins quickly dispersed at lower levels into the porphyry, and the lower rock could yield as little as .4 to 1 percent copper. Prior to the 1890s, such porphyry deposits were considered unprofitable and could not really be extracted by tunneling through the rock. The most profitable regional copper mine in the 1880s was the subsurface Copper Queen in Bisbee, Arizona, where the copper yield averaged 10 percent.[9]

The lack of purity in copper ore adds to the problem of how best to refine the metal for sale. Copper occurs in concert with a host of other metals, including silver, gold, lead, and iron. Throughout the borderlands it frequently was found embedded in limestone or quartz and alongside and within eruptive rocks such as porphyry. Because of this complexity, copper could not be refined in a single process. The goal in smelting copper was to produce a highly concentrated ore in a sulfide solution, called a matte, which contained as much of the copper and other metals (often silver and gold) as was possible to profitably separate from the ore. Waste material from this process, which frequently retained significant amounts of copper (almost always more than 2 percent) was called "slag" and dumped.[10] This matte was then further refined to separate the copper from the other metals. The expense of separating copper from the bulk of the ore through roasting, concentrating, or "ore dressing" and the two-part smelting process was a major hindrance to regional copper development well into the 1890s, despite the rising demand.

The Arizona Copper Company, backed by financiers in Edinburgh, Scotland, was one of the first companies in Arizona to turn a profit mining porphyry deposits. The company got its start not by sending out prospectors but by purchasing the Longfellow, Humboldt, Yavapai, and Detroit copper mines, all in the Clifton district in southern Arizona. At the time that the company began working the properties in 1883, there was little ore of a quality appropriate for smelting. Frustrated by the lack of high-quality ore

The rugged terrain of the borderlands posed major challenges in reaching mining and smelting sites. This train is headed for the industrial town of Morenci, Arizona. Underwood and Underwood, *Climbing the Last Loop, on the Mountain Railway to Morenci Copper Mines, Arizona*. Photograph retrieved from Library of Congress, https://www.loc.gov/item/2014649287/, Nov. 9, 2017.

in the ground, the head of smelting operations, a mining engineer named James Colquhoun, turned to the slag left by an earlier generation of miners who had worked the area using comparatively inefficient processes. He determined that there was enough copper in the slag to successfully mine and smelt it, after first concentrating the ore in primitive "jigs." The success of this method kept Arizona Copper in business, although without great profits, for about five additional years. By 1892, just as the company considered closing down, Colquhoun (by this time general manager of the company) devised a plan to concentrate some of the low-grade oxidized ores that the company's property contained, as Colquhoun himself recalled, "in plenty."[11] He proposed to do so using sulfuric acid produced on site from a small vein of iron pyrites.

Since Colquhoun's plan would require an expansion of the mine at a time when the owners were seriously considering closing it, the company brought in an outside mining engineer as a consultant—John Hays Hammond. After consulting Hammond, the board of directors determined to build the leaching plant that Colquhoun desired. Colquhoun himself traveled to New York with the company chairman to try to raise money for this new construction. The two men failed to find investors, which was not surprising, as Arizona Copper had a poor history as a dividend-paying operation and lacked any valuable property to put up as security. But Colquhoun persisted, finally persuading the board to allow him to draw on the profits from the company's rapidly diminishing ore reserves, while the board president persuaded the mortgage holder to forgo payments until the new leaching plant was operational.[12] As James Douglas observed, "Had not the shareholders [of the Arizona Copper Company] been willing to accept hopeful promises in lieu of dividends," the company would never have remained in operation.[13]

In order to produce enough ore to run through the leaching plant, Colquhoun engaged in "open quarrying," a method of extracting the ore from the top down. After scraping away the surface soil, the ore body was blasted open, and miners dug the ore in steps down through the ore body. This was an inexpensive way to extract both low-grade and first-class ores.[14] Colquhoun's combination of inexpensive mining and large-scale leaching facilities proved extremely lucrative for Arizona Copper, which by 1895 was able to pay both its mortgage and a small dividend to shareholders—a remarkable achievement for a company that was all but bankrupt three years earlier.

Beginning in 1896, Colquhoun also began working a couple of bodies of porphyry ores that the company had recently discovered containing 1 to 4 percent copper.[15] The pilot concentrating plant that Colquhoun devised could process one hundred tons of ore per day. Number 3, the first full-scale porphyry concentrator at Arizona Copper, came online in 1898 and could process 700 tons of ore per day. That year, Arizona Copper yielded over thirteen million pounds of copper—about 10 percent of the refined copper produced in Arizona—making it one of the top producers in the United States. By 1901, Arizona Copper was the eleventh largest producer of copper in the world.[16] Colquhoun himself pointed to some of these accomplishments in his modest personal history of the Arizona Copper Company to argue that he was the first mine engineer to successfully process porphyry copper, fully ten years before Daniel Jackling and the Utah Copper Company began mass mining it at Bingham Canyon in 1908.[17]

At approximately the same time that Colquhoun was experimenting with porphyry copper deposits at Clifton, Phelps Dodge was in the process of constructing its own mine to be excavated and processed on a mass scale.

The level of communication between Colquhoun at Arizona Copper and L. D. Ricketts at Phelps Dodge is impossible to determine, but as Ricketts worked as a consulting engineer for the Arizona Copper Company while holding a position as manager of mine operations at Phelps Dodge's latest acquisition at Nacozari, Sonora, it is almost certain that he was aware of the innovative concentration methods at Clifton. Although engineers at rival companies rarely collaborated outright, there was somewhat open communication between them. They read and regularly wrote about the processes they implemented in trade journals such as the *Engineering and Mining Journal* and the *Mining and Scientific Press*. Engineers also made regular site visits to nearby mines to talk to their professional colleagues, improve their knowledge of regional geology, and get a sense of the wealth of neighboring mines.[18] More casually, mining engineers were social peers and would often hear of new techniques or experiments as part of normal conversation.

L. D. Ricketts became the manager of the Moctezuma Copper Mine, the Phelps Dodge property near Nacozari, in 1897. Unlike Colquhoun, whose innovations at Arizona Copper were the result of necessary and expedient experimentation, Ricketts arrived at Moctezuma with a vision for the mining camp. Although Nacozari was famous for the wealth of its copper deposits in the 1860s, by the 1890s the area had a lot in common with other borderland copper mines—plenty of low-grade ore, some in the ground, some in slag heaps from earlier, less-efficient mining projects. Inspired by the mechanical processes he saw being implemented in factories, Ricketts envisioned Nacozari as a fully industrial mine, and he made it so.[19] The mine began operation in 1901.

The previous owners of the Moctezuma, the Guggenheim family, had attempted unsuccessfully to smelt the Nacozari ores on site. By installing a mechanized transportation process, however, Ricketts was able to remove significantly larger quantities of ore from the ground. He designed and built a much larger concentrator and smelter out of steel and connected the smelting works to the mining works with conveyor belts. A central electric power station powered the whole operation. Ricketts initially wanted the mine to run on gas, but its remoteness made that unfeasible, so from 1901 until the 1910s, the furnaces at Nacozari were fueled by wood from the surrounding forests. The size of the concentrator enabled the mine to process the larger quantities of ore required by borderlands copper mines. In its centralization of power, mechanization of infrastructure, and use of structural steel, Ricketts's mine at Nacozari was innovative and exemplified what the mass mines of the twentieth century would look like.

Blueprint of the reduction works at the Copper Queen, Bisbee, Arizona. James D. Hague Collection, folder T-18. Courtesy of The Huntington Library, San Marino, California.

Nacozari proved to be extremely profitable, and the enhanced technological system of the mine enabled still more centralization in the early twentieth century. From the time he took on the task of managing the mines at Nacozari, Ricketts worked to persuade James Douglas, by this time the head of all Phelps Dodge mining operations, to build a train to Nacozari to more readily connect the mine to its other regional operations. The standard-gauge line to Nacozari, a spur of the line that stretched from Bisbee to El Paso, was finally completed in 1904.[20]

In the years following Ricketts's transformation of the Nacozari mines, the landscape and economy of the border region transformed dramatically. In order to cope with the vast quantities of ore now being extracted from the ground, Phelps Dodge built a new smelter capable of processing over 1,500 tons of ore per day and a new settlement to serve the smelter, the border town of Douglas, Arizona. Although Phelps Dodge already ran one of the largest smelters in the country—the L. D. Ricketts-designed concentrator of the Detroit Copper Co., a Phelps Dodge subsidiary in nearby Morenci, the new smelter at Douglas had the capacity not only to refine copper from the many Phelps Dodge properties but also to work with ores from other mining companies along the border.[21] The technical and economic problem of how to treat low-grade ore in a cost-effective manner inspired, and perhaps demanded, the centralization of engineering work in the copper-mining industry. In turn, this changed the regional mining industry and the working lives of everyone associated with it.

Effect on Mining Engineers

Solving the problem of mining low-grade copper deposits had a profound effect on the organization of the working lives of mining engineers. In essence, mass-mining sites were extremely complicated building projects because low-grade copper had to be processed in enormous quantities. Such mines required a consulting engineer to suggest a strategy for digging and refining the ore and a managing engineer to oversee operations. In addition, an entire engineering department worked to determine where and how to dig, adopt or invent the best procedure for procuring and processing ores specific to the mine, and draw up and implement plans for linking the various parts of the operation mechanically.[22] The engineers who worked in the engineering department at Phelps Dodge after the methodological transition at Nacozari had a very different work experience from mining engineers of an earlier generation. Ricketts himself, for instance, worked throughout his career primarily as an independent contractor. Yet rather than receiving instructions from the owners or supervisors of the company, an engineer in an engineering department primarily interacted with other engineers, solving problems and designing cuts as directed by the head of the department. His work was subject to greater oversight by men who themselves possessed technical knowledge and the ability to effectively weigh the judgment of a mining engineer.[23]

The early career of Eugene Sawyer illustrates how changes within the profession might affect an individual career. Sawyer studied mining at Harvard and at the Lawrence Scientific School, where he was a middling student, receiving an A. B. from Harvard, followed in 1907 by an E. M.—a mining engineering degree—from Lawrence.[24] In 1910, he moved to Arizona to take a job in the engineering department of the Copper Queen. In Bisbee, Sawyer enjoyed all the privileges accorded to his status as a skilled white worker. He and his roommates, other mining engineers at the Copper Queen, lived in a house on Quality Hill.[25] Sawyer's work at the Copper Queen involved a certain amount of independent engineering and was not all office work. He spent a considerable amount of time underground touring the shafts, and when the company enacted a digging plan that he designed, he was gratified to be "mixing in with the proceedings."[26] Unlike mining engineers of previous generations who had done so much work alone in the office or in the field accompanied by only one or two assistants or fellow engineers, he also spent a great deal of time "mixing in" at the office with the other engineers. Fortunately for Sawyer, the company struck ore at two of the locations he recommended. He therefore impressed not only Walter Douglas, a son of James Douglas and managing director of the Copper Queen at the time, but also L. D. Ricketts, whom Douglas had hired as a consultant. Sawyer was awed by Ricketts, "the biggest mining man in this section of the country," and flattered by the senior engineer's humble manner. "[Ricketts] comes into my office nearly every day," the young engineer wrote to his mother, "[and] spends a good part of the afternoon sometimes over my maps and asks my opinion on nearly every question."[27] For an early career engineer, work in this engineering department threw him in with his professional peers on a daily basis and also brought him into contact with some of the biggest names in the profession.

As companies consolidated and there were fewer "new" mining prospects for engineers, they experienced a contraction in career options: it was increasingly difficult to obtain the kinds of experiences that Ricketts, or Louis Janin, or even James Douglas had early in life to establish a reputation and provide a good background for a consulting career. But young mining engineers also experienced gains from the increasingly automated, centralized, and industrialized operations. Sawyer, for instance, benefited enormously from joining an established profession. Although relatively less adventurous now, there were more positions available for such newly credentialed mining engineers. When on the job, young mining engineers were more likely to come into contact with senior engineers who could serve as mentors. Earlier mining engineers had entered a field in which work experiences were far more atomized and geographically dispersed, and more frequently than not had worked alone. Eugene Sawyer clearly profited from these new aspects of the profession: obtaining a good position shortly after graduating from Lawrence; working closely with L. D. Ricketts; and developing a close relationship with Walter Douglas. Of course, Sawyer was particularly fortunate in his position, as there were only a handful of engineers of Ricketts's stature in the United States. Yet while the specifics of his position were remarkable, the general trajectory of his early career was not.

Sawyer's eagerness to be involved, and his apparent talent as an engineer, helped him advance up the corporate ladder quickly. In August 1910, after only a few months in Arizona, Walter Douglas asked him to be the general manager for a new prospect that Phelps Dodge was considering in the

Catalina Mountains near the town of Oracle, just north of Tucson. Although pleased to be entrusted with the new responsibility, Sawyer was less enthusiastic about the work: "I was a little disappointed with the looks of the ground up there," he wrote. "If I had been doing [the survey], I don't know as I should have reported so strongly in favor of it as Grele [the surveying engineer] did."[28] Sawyer's work in Oracle, which consisted of surveying the ground at the mine site, purchasing mining equipment, transporting it into the mountains, and hiring labor, was not very different from that of James Douglas when Douglas first began operations in Arizona for Phelps Dodge. The path Sawyer took to that work, however, depended as much on his ability to work within the bureaucratic systems now in place at Phelps Dodge as on his talents as an engineer or skills as a manager.

When the mine at Oracle failed to turn a profit, Sawyer thought it likely that the company would move him back to the engineering office at Bisbee, although as a manager, rather than as a staff engineer. He had mixed feelings about this: the possible promotion "set me ahead in my profession more than five years," yet the work at Oracle had spoiled him. "It has been a wonderful experience" he reflected, ". . . [and] I like the exercise and all round work [of field engineering] best."[29] Despite his youth, he could likely have gotten recommendations from some of the top engineers in the business, had he decided to move on. Yet Sawyer did not consider leaving Phelps Dodge for a field-engineering position at a smaller company. This demonstrates a degree of company loyalty that is different from the attitude of Ricketts and his contemporaries, who changed company affiliations with alacrity and apparently little professional ill feeling. At the time when Sawyer met Ricketts, "the Doctor" was in fact salaried as general manager for the Cananea Consolidated Copper Company, owned by William Greene, and a consultant for Phelps Dodge, the local rival.[30] Sawyer, however, left his professional advancement in the hands of Phelps Dodge, determined to take the next position the company offered him.[31] After all, he noted, "everyone kow tows to the Queen," and he evidently concluded that the work in the engineering office at Bisbee was a better professional opportunity than looking elsewhere to find the "all round work" that he actually preferred.[32]

As mass mining became more entrenched in the copper industry and mining engineers were relatively less itinerant, their family lives changed as well. With few exceptions, nineteenth-century mining engineers tended to live apart from their wives and children. The incessant travel, long hours, and—particularly in the borderlands—lack of physical security and dearth of amenities in town did not make for an appealing lifestyle for the elite eastern women who were often the wives of mining engineers. Mary Halleck Foote, for instance, who had a successful career herself as an author and illustrator of highly romanticized tales of life on the frontier and in the mines of California, Idaho, and Mexico, relished the extent to which she flouted convention by following her mining-engineer husband all over the western hemisphere. Although she occasionally socialized with the wives of other engineers, she did not maintain a home in San Francisco or New York through the 1870s and 1880s, as she and her husband could not afford to maintain two residences.

Engineers worried about working in places that were difficult to reach. When discussing whether he would take over the on-the-ground management of the Cusihuiriachic mining company in 1886, Ellsworth Daggett explained, "If the inducements are sufficient and the locality such that I can take my wife I would go for a year or some fixed time." For Daggett, being on site was only a possibility if it were for a short period of time. While Mrs. Daggett seems to have been willing to live at a mine site under certain very specific circumstances, it is also possible that Daggett was not paid enough to maintain two separate households.[33]

William F. Staunton met his wife, Mary Neal, in Tombstone while he was working at the Tombstone Mining and Milling Company in the 1880s. Mary's sister, Annie Cheyney, was married to the company's managing director. Although most mining engineers met their wives while in school or when visiting friends and families back East, the trajectory of the Staunton's married life together was probably fairly typical. After their marriage, the Stauntons lived with the Cheyneys in a two-story company-owned adobe near the mines in Tombstone. The sisters apparently spent most of their time together, engaged in a variety of artistic pursuits that their husbands regarded with indulgence if not admiration.

When Staunton moved to superintend the Congress Mine, near Prescott, Arizona, his wife and son went with him, although the question of where Mrs. Staunton was to live was a difficult one. The manager of the Congress lived in Prescott with his wife in a house that was not large enough for two families, so Mrs. Staunton moved to Congress itself. After a few years, other workers brought their wives and children, and the mine had a small community of families of which the Stauntons were the most prominent. After moving within Arizona a couple more times, Mrs. Staunton and the children eventually moved to southern California.[34] The great distance between Staunton's work in Arizona and his family was doubtless ameliorated by both the relative ease of train travel by the 1910s and the Stauntons' rising wealth.

The reorganization of the mining business, however, gave more engineers the opportunity to consider the possibility of bringing their families to the places they would be working. At one time, the ability to send the family to live in Santa Barbara or Los Angeles was an indication of the success of a mining engineer. By the early twentieth century, the opportunity to work at a location where one might bring a wife, and possibly children, proved important to many mining engineers when considering which job to take. "I wish to take my wife to a district less isolated than this [the district he was currently in]," one engineer wrote in an application letter, "and one where there may be more chances of advancement."[35] Certainly this engineer sought a better opportunity for himself, but his primary impulse in seeking new work was to move his family to a larger mining camp. Another engineer, hoping to relocate from Alaska to Jerome, Arizona, was more explicit about his wife's needs.

> I am writing fully not because I am dissatisfied with my present position or its opportunities, but because the long Alaskan winters and great distance from other large mine operations and from communication with the states are conditions which preclude the possibility of an indefinite stay in this district with my wife and baby.[36]

This engineer clearly did not think that being explicit about his wife's needs would in any way diminish his own chances of getting hired. Indeed, he may have had reason to believe that being married would make him a more attractive employee. Employers who had a vested interest in the job satisfaction of their most expensive workers cared about the familial status of mining engineers. Marriage was a marker that denoted a man's seriousness of purpose and likelihood of remaining on the job. One man recommended a mining engineer to the manager of the Cananea Consolidated Copper Company by explaining that the engineer, Charles Ratcliff, wanted a new challenge: "P. S. [:] Ratcliff is a married man, about 38 years old, and has a small family, I believe, who are with him in San Pedro."[37] While Mary Halleck Foote, as an educated white woman, had been such an aberration in the mining community that her "salons" in remote locations such as Leadville and Idaho were famous among the early generation of mining engineers, by the early twentieth century, the upper-middle-class wives of mining engineers were frequently found in the larger mining camps.

With the shift to mass mining, the expertise needed by mining engineers also underwent a change. Nineteenth-century mining engineers spent a great deal of their time conducting surveys, assaying ore bodies, and constructing plans of operation and the technologies that would enable those plans to succeed. With the expansion of the engineering department at ever-larger mine sites, those particular skills were practiced by fewer and fewer men. Rather, in addition to geology and surveying, engineers needed to know more about metallurgy and chemistry than ever before and to have a solid understanding of hydraulics and steam engineering.[38] After graduation, mining engineers needed to stay informed of the latest developments in mining techniques, which began to change at a rapid rate. The membership of professional societies surged, filled with engineers who wished to keep up with the new sciences, and new, more regional societies were formed, often explicitly to share scientific and technical knowledge among their members. What mining engineers actually did encompassed a broad range of administrative and managerial tasks, far greater than might be supposed from reading the articles in mining journals.[39] In their most ideal state, the professional organizations provided a forum for face-to-face meetings between mining engineers and facilitated discussion within the community about how to approach these new tasks.

The larger and older organizations, most notably the American Institute of Mining Engineers, remained critical as initiators of major networking events. In 1901, led by its president Eben Olcott, who had spent years as a young engineer working in Sonora, AIME sponsored a trip to Mexico. The ostensible mission was to forge strong ties between Mexican and American engineers. As the *Engineering and Mining Journal* put it, "Mexican engineers surely have lots to learn from U.S. mining." In reality, of course, the trip functioned as a networking opportunity for the American engineers, 160 of whom gathered in two specially chartered Pullman trains, fully equipped with sleeping compartments, multiple drawing rooms, and two private cars, one of which was for James Douglas, president of Phelps Dodge. The mining engineers traveled together from New York to Chihuahua, stopping at a number of mines both large and small, as well as at the new ASARCO [later Guggenheim]-owned smelter in Aguascalientes.[40] The American engineers were impressed by the developments they saw and also by what they learned of mining conditions in Mexico, where (they were told) there were no labor strikes, and laws were more favorable to mining. The law of the apex, for instance, the bane of mining engineers throughout the United States, was not in effect in Mexico, and therefore the number of boundary disputes in Mexico was significantly smaller than in the U.S. The trip, which to modern eyes is indistinguishable from a corporate junket, concluded with a dinner with President Porfirio Díaz, about whom *Engineering and Mining* gushed,

"This opportunity to see and grasp the hand of one of the greatest men in modern history was fully appreciated by all."[41]

Smaller, more regional organizations than the AIME also played a critical role connecting mining engineers to one another for social and professional reasons. One example is the Society of Engineers and Metallurgists of the Republic of Mexico, established about 1907. Membership numbers are difficult to determine, but the Society's pamphlets, published in English and in Spanish, contain essays by several English-speaking engineers in Mexico City and Chihuahua, suggesting that the group's writings circulated at least throughout northern Mexico and the capital and that working mine engineers were interested in supporting such an organization. The avowed aims of the Society were to promote the exchange of professional expertise among members; to investigate matters "having some bearing...upon the condition of the mining industry of Mexico, or the professional workers associated with it"; and more generally to liaise between the community of mining engineers and the Mexican government, so that mining engineers could present a united front in a nation that "stands in the very front rack in a comparison with the other mining countries in the world." As one pamphlet explained, the cooperation between members of the Society would address instances

> in which two or more men are working on the same problem, though in districts several hundreds of miles apart. One may find a ready solution...but the others may be less fortunate, and being in ignorance of the fact that there is a solution they may finally give the fight up in despair, or their work may result in dire failure.[42]

The stated goals point to a persistent concern within the community of mining engineers that as a profession they be considered the honest, and final, arbiters of mining valuation and practice.

The mass-mining techniques that spurred these professional and social rearrangements throughout mine engineering occurred at the turn of the twentieth century, as engineering institutes were turning out more and more mining graduates.[43] Such changes within mine engineering also stripped the profession of those functions that had made mining engineers into romantic figures in the nineteenth century. In general, they no longer ventured into unknown territory as part of exploring parties, as Louis Janin had done in the 1860s with the Butterfield Expedition. Nor were they forced to live in relative isolation, on contested territory and without the protection of the U.S. government, as Raphael Pumpelly had in Tubac. They no longer had to prospect in the open countryside on horseback for days, searching for outcroppings that suggested the presence of a large deposit of gold, silver, or copper. When mining engineers did undertake relatively isolated work, as did Eugene Sawyer, it was under the auspices of a large company, with ample capital investment and a relative ease of communication with the outside world. Although it still took Sawyer two weeks to haul a boiler from Tucson to Oracle, he himself was able to ride into Tucson, twenty miles away, with great frequency (although he was somewhat dependent on the stagecoach schedule), and Oracle also had a phone line to Tucson.[44]

Despite these changes, in the early 1900s the public still perceived a mining engineer's life as one of romantic adventure. In part, this is because of the social and political prominence of a handful of mining engineers such as John Hays Hammond or Herbert Hoover. Hammond, who spent a considerable amount of time in the Southwest and northern Mexico, was a heroic figure in the eyes of many, on account of his exploits in Africa with Cecil Rhodes. Hoover, who was to rise to worldwide prominence and domestic popularity during World War I for his role organizing the relief effort for Belgium, was already well known in mining circles as a young engineer who made a fortune mining in Australia and working in China just before the Boxer Rebellion.[45]

Another indication that the cultural status of mining engineers remained firmly embedded in the world of adventure is the relative popularity of H. Irving Hancock's series of action books for boys, The Young Engineers, published in the 1910s. With titles such as *The Young Engineers in Mexico; or, Fighting the Mine Swindlers*, and *The Young Engineers in Arizona; or, Laying Tracks in the Man-Killer Quicksand*, these books chronicled the adventures of two young men who gained experience after high school working as surveyors in diverse and exotic locations out West.[46] The young heroes started out as civil engineers working on railroad projects, working their way into the unknown deserts, caverns, and mountains of the southwestern U.S. and northern Mexico, while amassing technical experience and frontier credibility. In the mythology of the series, Tom and Harry got involved in a gold mine that went into bonanza (*The Young Engineers in Nevada; or, Seeking Fortune on the Turn of a Pick*) and, having learned from that experience all there was to know about mining, were full-fledged consulting mine engineers by the time they uncovered a tremendous mining fraud in *The Young Engineers in Mexico*. Although stylized juvenile fiction, the Young Engineers series nonetheless encapsulates many of the tropes of mine engineering first expressed by both travel writers such as journalist J. Ross Browne and fiction writers such as Mary Halleck Foote, who wrote about western mining between the 1870s and the 1890s.

Although this romance persisted in the popular imagination, and at least one mining engineer admitted that it was a key component of his decision to pursue the profession, it was at odds with how a number of mining engineers, both elite and rank-and-file, were beginning to think about their careers.[47] Increasingly, engineers observed a distinct lack of opportunity for younger colleagues to gain access to the kinds of consulting work that could prove both interesting and lucrative. When Theodore Roosevelt solicited the advice of his old friend, John Greenway, president of the Arizona Copper Company, as to whether his son Kermit should pursue a mining degree, Greenway responded with caution. To succeed, Greenway noted, Kermit did need a degree, as "entering the business Kermit with a Mining Engineer's Degree....Will have a better chance to win out than Kermit without the training such a Degree means." Clearly, Greenway believed that all the connections in the world could not help a man who wished to advance in the profession without a degree.[48] However, Greenway strongly suggested that before entering university Kermit spend a year in underground work to get a feel for what mining entailed. He offered to take Kermit on himself or to recommend Kermit to "our mutual friend, Cleveland Dodge [of Phelps Dodge]," who ran the Copper Queen Mine. What Greenway somewhat equivocally states to Roosevelt regarding the significance of a mine engineering degree was in fact becoming a hard-and-fast rule.

In the early twentieth century, a young mining engineer, whether in Arizona or Sonora, had a very different career trajectory than did a mining engineer starting out just a generation earlier. One manager at a Phelps Dodge property kept a journal clipping that addressed this circumstance. It noted,

> Frequent attempts have been made to arrive at a satisfactory definition of an engineer with more or less varied success. An engineer is one who, through application of his knowledge of mathematics, the physical and biological sciences, and economics, and with aid, further, from results obtained through observation, experiences, scientific discovery, and invention, so utilizes the materials and directs the forces of nature that they are made to operate to the benefit of society. An engineer differs from a technologist in that he must concern himself with the organizational, economic, and managerial aspects, as well as the technical aspects of his work.[49]

The alleged scope of an engineer's realm speaks to some extent to the importance that mining engineers—justifiably by this point—placed on their own work within the mining industry. The introduction of the mass-mining techniques of mechanized transportation and exceptionally large smelting facilities required knowledge of math and physics. Establishing operations to treat the low-grade borderlands copper ore required knowledge of chemistry. Adapting these processing techniques to handle the range of ores processed in centralized treatment plants in places such as Douglas required a flexible inventiveness. And drawing a budget for all this rather expensive work and still manage to pay dividends to investors required a basic understanding of economics. Although the breadth of work that a single engineer undertook was no greater than in previous generations (and in many cases was much narrower), no engineer could do his job unless he worked closely with the engineers who oversaw or designed other aspects of the mine and thoroughly understood their functions and the pressures under which they operated. That this Phelps Dodge manager kept a definition of "engineer" in among his notebooks is indicative of the transition underway within the profession of mine engineering.

The early twentieth century is the time historian Edwin Layton identifies with "the revolt of the engineers," a moment when certain branches of the engineering profession recoiled from over-identification with capitalists and embraced the rhetoric of the Progressive reform movements. Layton's engineers recognized that as technical workers, engineers were uniquely placed to be reform-oriented political actors. He points out, however, that the "reform" attempts of the AIME ended before World War I and chiefly consisted of amending the membership rules to be friendlier to business.[50] This change reflects the new subservience of mining engineers to the needs of the large corporations. Mining companies had become powerful local entities in economic matters—recall Eugene Sawyer's succinct summation "everyone kow tows to the Queen." The mass mining companies were also just as powerful in the lives of up-and-coming engineers, who for the sake of professional and personal stability sought positions in successful mining companies. The mining engineers of the early twentieth century were a far cry from the "westering" engineers who headed into the silver mines of Tubac, or Alamos, forty years earlier. The mining engineer's "dilemma," as Layton observed, had become "at base, bureaucracy, not capitalism."[51]

Notes

1. *Engineering and Mining Journal*, 95 (March 8, 1913):534; cited in Hovis and Mouat, *Miners*, 455.
2. Hovis and Mouat, 455.
3. Proponents of gas drilling in the Marcellus Shale in the eastern U.S. in the

2010s made a similar argument about the financial feasibility of hydraulic fracturing—an argument that is somewhat less persuasive than the argument for low-grade copper in 1900.

4. The classic text on "sociotechnical systems" and the embeddedness of technology within society is Thomas P. Hughes, *Networks of Power: Electrification in Western Society, 1880–1930* (Baltimore, MD: Johns Hopkins University Press, 1983); see also Bruce Bimber, "The Three Faces of Technological Determinism," in Merritt Roe Smith and Leo Marx, eds., *Does Technology Drive History?* (Cambridge, MA: MIT Press, 1994); Wiebe Bijker and Trevor Pinch, "The Social Construction of Technology," in Wiebe Bijker, Thomas Hughes, and Trevor Pinch, eds., *The Social Construction of Technological Systems: New Directions in the Sociology and History of Technology* (Cambridge. MA: MIT Press, 1987); Ronald R. Kline, *Consumers in the Country: Technology and Social Change in Rural America* (Baltimore, MD: Johns Hopkins University Press, 2000); Nye, *Electrifying America.*

5. U.S. Bureau of Mines, *The Production of Copper in 1910* (Washington, D.C.: Government Printing Office), 23; Michel E. Parrish, *Mexican Workers, Progressives, and Copper: The Failure of Industrial Democracy during the Wilson Years* (La Jolla, CA: Chicano Research Publications, 1979), 4–5. Historian Mark Wyman argues that despite the militancy of the WFM through the 1890s, by 1910 or so the miners' unions were relatively weak, in part because the "machinery, technology, crews, systems of ownership, and dynamics of local politics" were constantly shifting in the metals-mining industry, forcing unions to perpetually change their approach to the problems of fair labor; Wyman, *Hard Rock Epic*, 256–258.

6. LeCain, *Mass Destruction*, 38–39; 47; 130–131. LeCain is part of a new movement of so-called "envirotech" scholars who are extending the sociotechnical realm to include the effect of technological systems on the natural world. See, for example, Frehner, *Finding Oil*; Chris Jones, "A Landscape of Energy Abundance: Anthracite Coal Canals and the Roots of American Fossil Fuel Dependence, 1820–1860," *Environmental History* 15, no. 3 (2010): 449–484; Sara Pritchard, *Confluence: The Nature of Technology and the Remaking of the Rhône* (Cambridge, MA: Harvard University Press, 2011).

7. For instance, a 1915 pamphlet praises Jackling, along with two other mining engineers, for pioneering the work on porphyry copper, without mentioning James Colquhoun, who is discussed in detail below. [Pamphlet], Horace C. Baker, "Copper" (New York: Charles A. Stonehame, 1915), folder 5, box 8, HAT.

8. Santa Rita, New Mexico, about fifteen miles east of Silver City, should not be confused with the Santa Rita silver mining district near Tucson in the Arizona Territory discussed in chapter 1. Native ore formed the bulk of the copper present in the Lake Superior copper district in Michigan's Upper Peninsula, which, until the twentieth century was one of the most productive mining areas in the world. Edward Dyer Peters, *Modern Copper Smelting*, 7th ed. (New York: Scientific Publishing Co., 1895), 7.

9. J. Foster and J. D. Whitney, *Report on the Geology and Topography of a Portion of the Lake Superior Land District*, part 1 (Washington, D.C.: House of Representatives, 1850), 177; Peters, *Modern Copper Smelting*, 10, 21–24; James Douglas, "Conservation of Natural Resources," *Transactions* AIME 40 (1909), 422.

10. Peters, *Modern Copper Smelting*, 234. For more on the Copper Queen in the 1880s, see chapter 4.

11. James Colquhoun, *The Story of the Birth of the Porphyry Coppers* (London: William Clowes, [1931]), 8–10.

12. Colquhoun, *Birth of the Porphyry Coppers*, 12.

13. Douglas, "Conservation of Natural Resources," 424.

14. Colquhoun, *Birth of the Porphyry Coppers*, 12–14; Henry Stafford Osbourne, *A Practical Manual of Minerals, Mines, and Mining*, 2nd ed., rev. (Philadelphia: Henry Carey Baird, 1896), 300–303; Horace J. Stevens, ed., *The Copper Handbook*, vol. 3 (Houghton, MI: Horace J. Stevens, 1903), 183.

15. Stevens, *The Copper Handbook*, 183.

16. D. Houston and Co., *Copper Manual: Copper Mines, Copper Statistics, Copper Shares* (New York: D. Houston, 1899), 286; Horace J. Stevens, ed., *Mines Register: A Successor to the Mines Handbook and the Copper Handbook*, vol. 3 (Houghton, MI: Horace J. Stevens, 1902), 566–568.

17. Note that in *Birth of the Porphyry Coppers* Colquhoun does not dispute that the Bingham Canyon Mine in Utah operated with vast efficiencies of scale that Arizona Copper never reached and that the overall method of mining at Bingham and its open-pit progeny was vastly different from that practiced by Colquhoun in Clifton-Morenci in the 1890s and early 1900s. Colquhoun, *Birth of the Porphyry Coppers*, 20, 23–27.

18. As discussed in chapter 3.

19. Truett, *Fugitive Landscapes*, 56; Cleland, *A History of Phelps Dodge*, 130–131; Press Reference Library (Western Edition), *Being the Portraits and Biographies of the Progressive Men of the West*, vol. 2 (New York: International News Service, 1915), 123; Rickard, *Interviews*, 436–437.

20. Rickard, *Interviews*, 436–438.

21. Stevens, *The Copper Handbook*, 278, 289; [Douglas Biography], 190–191, 208–209, 219, folder 3 box 1, JD; Truett, *Fugitive Landscapes*, 83; Cleland, *A History of Phelps Dodge*, 117, 125, 133–135.

22. Hovis and Mouat, "Miners," 441.

23. Robert Peele, *Mining Engineer's Handbook*, vol. 1 (New York: Wiley, 1918), 1269.

24. Records of Students in the Lawrence Scientific School, 1888–1911, Harvard University Archives, Cambridge, MA.

25. Eugene Sawyer to Mrs. Sawyer, January 10, 1910 and January 18, 1910, folder 1, box 1, MS 360 Eugene Sawyer Papers, Arizona Historical Society, Tucson, AZ (hereafter ES).

26. Eugene Sawyer to Mrs. Sawyer, March 14, March 22, and August 7, 1910, folder 1, box 1, ES.

27. Eugene Sawyer to Mrs. Sawyer, August 7, 1910, folder 1, box 1, ES.

28. Eugene Sawyer to Mrs. Sawyer, August 7, August 15, and August 22, 1920, folder 1, box 1, ES.

29. Eugene Sawyer to Mrs. Sawyer, March 17, 1912, folder 3, box 1, ES.

30. Prior to accepting the salaried position at Cananea, Ricketts had already turned down the opportunity to serve as director of Greene's company due to having too much work of his own and no time or patience for working with a large board of directors, "many of [whom] have little practical knowledge of the mining business and very few technical knowledge of that business. Under these circumstances I feel that I could not render you the service I...could give you if I served on a small board made up of men thoroughly familiar with the copper mining business." Louis D. Ricketts to William C. Greene, May 4, 1904, folder 3, box 14, LD.

31. Eugene Sawyer to Mrs. Sawyer, July 23, 1912, folder 3, box 1, ES.

32. Eugene Sawyer to Mrs. Sawyer, September 25, 1910, folder 1, box 1, ES.

33. Ellsworth Daggett to James D. Hague, December 6, 1885, folder M-13, JDH.

34. Staunton, "Memoirs," 71–79, 102–152, WFS.

35. Allan D. Robinson to James E. Hyslop, August 5, 1909, folder 1, box 3, H-B. See also Cyrus Cole to David Cole, January 23 and November 29, 1903, folder 2, box 1, Cananea Mining Papers, 1893–1930, MS 1311, Arizona Historical Society; G. J. Geer Jr., to John Greenway, November 9, 1912, folder 2720, box 196, Greenway Collection; Staunton, "Memoirs," 102, WFS.

36. Henry DeWitt Smith to Robert E. Tally, April 19, 1917, folder 1, box 1, Henry DeWitt Smith Collection, #7721, American Heritage Center.

37. J. O'Grady to George Mitchell, October 31, 1901, Cananea Consolidated Copper Company Records, 1898–1969, MS 1032, Arizona Historical Society.

38. Engineering schools changed their course of study to reflect these new professional realities. See Governing Board Minutes, folder 56, box 6, RU 819.

39. Ochs, "The Rise of American Mining Engineers," 278–79.

40. *Engineering and Mining Journal* 72 (November 30, 1901), 693–698.

41. Ibid., 695.

42. [unknown], "Aims and objects of the society of Engineers and Metallurgists of the Republic of Mexico," typed ms. [1907], and H. S. Denny, "Aims and Objects of the Society of Engineers and Metallurgists of the Republic of Mexico," [1908–1909], FF13, H-B.

43. See chapter 2.

44. Eugene Sawyer to Mrs. Sawyer, August 22 and September 5, 1910, folder 1, box 1, ES.

45. Jeremy Mouat and Ian Phimister note that Hoover's fame, among mining engineers at least, stemmed from his financial acumen rather than his skill as an engineer or his heroism under duress. Hoover's fortune dates from his work in China. During this time, he was implicated in a nasty financial scandal, accused of defrauding Chinese mine owners during the confusion surrounding the Rebellion. Hoover himself disliked talking about his time in China. Mouat and Phimister, "The Engineering of Herbert Hoover," *Pacific Historical Review* 77, no. 4 (2008):558.

46. H. Irving Hancock, *The Young Engineers in Mexico* (Philadelphia: Altemus, 1913); *The Young Engineers in Arizona* (Philadelphia: Altemus, 1912); *The Young Engineers in Colorado* (Philadelphia: Altemus, 1912); *The Young Engineers in Nevada* (Philadelphia: Altemus, 1913).

47. Fred Bailey Oral Interview, no. 115, Oral History Collection, University of Texas, El Paso.

48. John Greenway to Theodore Roosevelt, November 29, 1911, folder 2718, box 196, Greenway Collection.

49. "What Is an Engineer?" box 16, Frank Ayer Collection, #3341, American Heritage Center.

50. Layton, *Revolt of the Engineers*, ch. 2. See also Ronald Kline, "From Progressivism to Engineering Studies: Edwin T. Layton's *Revolt of the Engineers* Reconsidered," *Technology and Culture* 49, no. 4 (2008):1018–1024.

51. Layton, *Revolt of the Engineers*, 1.

CHAPTER SIX

LEGIBILITY AND THE TECHNOCRATIC LANDSCAPE

By the early twentieth century, mining engineers were crucial to the development and expansion of mass mining by the now large and powerful U.S. mining companies in the borderlands. Such corporations were not, however, the only companies for which mining engineers worked. Mining engineers were also employed by development, or exploration, enterprises, two kinds of mining investment companies that arose in the late nineteenth century and operated in a distinctly different way than did traditionally established mining companies, viewing the mining industry and the mining landscape through a broader lens than just the extraction and processing for market of a single mineral. Development and exploration companies had an expansive, regional, and in some cases, global, perspective and relied on mining engineers to map that region and to act on the promise of the resources they highlighted. Usually formed as joint-stock operations, for the decades surrounding the turn of the century they were important intermediaries in bringing capital to the mining frontier.[1]

For a company such as Phelps Dodge, all decisions revolved around the primary mining sites. All of L. D. Ricketts's innovations in Nacozari and Douglas, for instance, were designed to increase the efficiency and productivity of the company's copper-mining project. In buying property, as in Nacozari, the company looked for copper mines; in constructing equipment, as at Douglas, the company looked to maximize its ability to process ore from different locations. The goal of Phelps Dodge was to process copper for the market. By contrast, most companies with "development" or "exploration" in their company names sought to find underperforming or abandoned, ignored, or poorly managed mines and to rehabilitate them using the latest extraction and processing technologies and engineering methods. Alternately, a company's operators might act as agents for an interested investor or group of investors rather than as developers in their own right. Development and exploration companies took a broad view of an entire region, considered the different types of minerals present, and strategized how to create financial returns. These companies viewed the mining landscape with a level of abstraction that was quite different from that of more traditional mining companies—even the increasingly large mass-mining corporations—and were dependent upon the professional credibility and technical expertise of mining engineers.

Mining development companies occupy an important niche in the expansion of the United States at the turn of the twentieth century. The massive expansion of the U.S. economy during these years could be considered the result of a concerted push among policy-makers and intellectuals to seek new markets for American goods and services.[2] Yet mining development companies operated on a different model, bringing greater efficiency to an extant extractive industry by investing in infrastructure and exploiting local labor and industrial resources. Connecting financiers and industrial technology to aggressive infrastructure development, development companies, no less than the more traditionally structured mining corporations, helped to turn mining into a truly transnational industry, linking labor, expertise, finance, and profits across national lines.[3] They did so by treating the mining districts of the Southwest and Mexico as a coherent "technocratic landscape," rather than as a set of discrete sites in competition for resources and market share.

Technocracy is a twentieth-century term used to describe a society or organization controlled by a group of technical experts. By the 1890s, the mining industry had been transformed into a business fully dependent on mining engineers, who had supplanted apprentice-trained and self-taught technical workers as the significant authoritative voices within the industry. The writer and intellectual most associated with the concept of technocracy, Thorstein Veblen, was entranced by the idea that the rationalist and technical authority of engineers could transform the fabric of society. He also believed engineers could "size up men and things in terms of tangible performance, without commercial afterthought."[4] Although his faith in the radicalism of engineers was misplaced, Veblen had a thorough understanding of the importance of systematic thinking as a part of an engineer's technical training.[5] By virtue of the spatial, logistical, and (ironically, given Veblen's philosophy) commercial implications of their work, mining engineers of the late nineteenth and early twentieth centuries exemplified technocratic thought. Although mining companies were cautious about sharing the details of their production with rivals, mining engineers were acutely aware that in their business, knowledge was power. The more he understood about regional similarities and deviations in ore presentation and processing techniques, the more efficiently a mining engineer could work his own property. Mining engineers sought global technical and geological knowledge of the region in which they worked, and they worked and conceptualized the borderlands as a landscape governed by technocratic concerns.

Development companies were not always large and bureaucratic. They could just as easily be small, formally organized associations of friends and acquaintances. J. M. Seibert, an agent dealing in mining stocks for Wells Fargo in Mexico City, for instance, spent his years in Mexico seeking additional income from mining and ranching investments. After losing money in a coffee farm, he explained to a friend, "There are so many attractive mining schemes in this country [that] there is absolutely no use trying to get any one... [in Mexico] to go into anything in the United States that he knows nothing about."[6] Seibert, as a mining stock agent, knew a great deal about mining in Mexico. Yet despite his solid grasp of the industry, Seibert was not himself a technical man; all his ventures were mediated through the work and reports of mining engineers. Indeed, because of Seibert's position within the industry, he was able to make use of the reports and opinions of mining engineers simply by talking to and corresponding with them, for he was in no position to pay for an expert himself.

In the fall of 1901, Seibert and his associates determined to rethink their investment strategy, following years of disappointment in ranching ventures. They did so by trying to form an organization called the International Land, Mining, and Exploration Company, with capital stock of $100,000 and the majority of the ownership to go to three capitalists in Texas. Seibert's role was to provide access to mining expertise. The intended purpose of this company was to "obtain options of promising mine claims and endeavor to find buyers."[7]

Getting the company off the ground proved to be a challenge for Seibert, who soon decided that the land company needed to hire its own mining engineer to "investigate and report on properties that may be offered."[8] Hiring their own mining engineer, however, rather than contracting one on an ad hoc basis, required even more capital funding, and so Seibert and his colleagues spent considerable time attempting to raise money, especially after their relationship with the three Texans collapsed. Seibert himself was so strapped for cash that he had to borrow money in order to invest in his own exploration company.[9] The company limped on, nonetheless, and over a couple of years did some actual work: consulting with mining engineers and other mining investors over properties; visiting likely prospects; and making small investments in projects managed by other companies in the hopes that they would pay off and enable Seibert and his associates to make a substantial investment in a property of their own.[10]

Mining engineers were the critical actors in development companies; without them the companies could not function. However, the position of mining engineers within such companies varied dramatically. These distinctive roles illuminate an important shift in the role of mining engineers throughout borderlands mining districts: engineers changed from the instigators of exploitation they had been in the mid-nineteenth century into enablers of broad-based corporate expansion. A brief examination of the work of mining engineers in a development company and in a large exploration company illustrates the central importance of technical workers in these diversified organizations. It further illuminates how a technocratic landscape was established through the mining districts of the southwestern United States and northern Mexico.

The Development Company of America

When William F. Staunton was growing up in Ohio, the example of his neighbor, John A. Church, a mining engineer who worked in the Arizona Territory, inspired him to study engineering. After obtaining a degree in mine engineering from the Columbia School of Mines in the early 1880s and gaining varied experience in east-coast mining ventures, Staunton ventured out West, finally settling in Tombstone with the assistance of his erstwhile neighbor, now an established and well-respected engineer working the mines of the Tombstone Mill and Mining Company.[11] Tombstone in 1883 was a boomtown and appeared to the young engineer to be a mining camp with a future. Unfortunately for Staunton, appearances were deceiving. Tombstone was peaking; the mines tapped out almost as quickly as they boomed. By the time Staunton was promoted to superintendent, in 1887, the company was mired in debt by the rapidly dropping value of silver on the American market, and Staunton was forced to close the smelter and ship the small quantities of ore the mines produced off site for processing, a move that dramatically reduced the likelihood that the company would generate dividends. At around the same time, the Tombstone miners struck water just below the 500-foot level. For a few years, several companies, including the Tombstone Mill and Mining Company, struggled to drain the basin under the mines, but a fire at the Contention Mine destroyed all the most valuable pumping equipment and with it the ability to mine at depth.[12] The rising water levels in the Tombstone mines, the weakening value of silver, and the failure of the pumping system combined to end the project of mining ore in Tombstone.

Around the same time as the Contention fire, Staunton was offered the position of superintendent at the Congress Mine, a gold and silver mine in Yavapai County near Prescott, Arizona. The Congress was successful in all the ways the Tombstone Mill and Mining Company was not, producing $5 million in gold profits through the 1890s. By 1900, Staunton was a wealthy man, superintendent and part owner of one of the most profitable precious metals mines in the region.[13]

One of the owners of the Congress Mine was Frank Murphy, who has been described by one historian as "the Southwest's greatest financier." The brother of the governor of territorial Arizona, Nathan Oakes Murphy, Frank Murphy purchased his first mining claim in Prescott in the 1880s. He used profits from the sale of that property to establish himself as a player in southwestern finance and had a hand in funding many infrastructure projects in Arizona and northern Mexico, particularly railroads. Murphy was the primary developer of the Santa Fe, Prescott, and Phoenix Railroad (SFP&P), and the El Paso & Southwestern Railroad short line to Tombstone. He also eventually negotiated the sale of the SFP&P to the Atchison, Topeka, & Santa Fe (ATSF) and of the Phoenix and Eastern to E. H. Harriman of the Southern Pacific, connecting significant southwestern mining sites to major transcontinental rail operators and thus to the rest of the country. Murphy's interest in the Congress Mine was almost certainly the reason that, despite its remote location at the foot of the mountains west of Prescott, the Congress was connected to a rail line early in its operation.[14]

In 1901, Murphy and a few other men involved in the Congress, including Staunton; the president of the Congress Mine, E.B. Gage; the lawyer Henry Robinson; and Benjamin Cheyney, a Boston financier and part-owner of the SFP&P, formed the Development Company of America (DCA), a corporation devised to run development schemes in the Southwest, principally Arizona, and funded via the public sale of 6 percent bonds. Frank Murphy was the vice president and maintained an active managerial interest in the operation. The initial investors retained a 51 percent ownership stake in the DCA and held 20 percent of the company's dividends as an operating budget.

The Development Company's first action was to purchase the Congress Mine and all the mines in the towns of Tombstone and Poland, Arizona. The DCA later purchased the Silverbell Mines of the Imperial Copper Company near Tombstone and in conjunction with the Southern Pacific built a rail line from Silverbell to the new Phelps Dodge smelter at Douglas. By 1906, the DCA was the largest holding company in the Southwest and ranked as the seventy-sixth largest company in the United States, with capital assets

The scale of works at Tombstone in the late nineteenth and early twentieth century can be gathered from this view of a hoist and ore dumps, which foreground the cords of wood necessary to keep the steam-powered machinery functioning. Image PN 137503. Courtesy of The Huntington Library, San Marino, California.

in excess of $34.4 million.[15] Staunton, who continued to superintend the Congress, was also given responsibility for the management of mining Tombstone, where the DCA united all the mines under the aegis of the Tombstone Consolidated Mining Company. He soon also assumed the superintendency of the Silverbell.[16] Staunton's role in the company was technical manager, overseeing the engineering and mining decisions of three fairly disparate mining operations.

The DCA's expansion at this time was qualitatively different from the contemporaneous expansion of Phelps Dodge. Where the growth at Phelps Dodge was driven by technical and engineering problems—processing ore onsite at Nacozari was expensive, for instance, but could be made profitable by mixing local ores with those from Bisbee, and processing at Douglas—the expansion of the DCA was an end in itself. As a development company, its resources were necessarily diversified, and the acquisition of new property was driven by the desire always to be able to offset losses, rather than the need to streamline a technological system.

The fate of the mines at Tombstone, and of William Field Staunton as the superintendent of DCA mining operations, exemplifies the position of mining engineers such as Staunton in a diversified development company such as the DCA. Despite the failure of the Tombstone mines in the late 1880s, they were generally thought by mining engineers to contain plenty of silver and gold, albeit in quantities requiring modern methods and great patience to extract.[17] The difficulty of working the Tombstone mines on a mass scale became apparent after the Tombstone Consolidated took over production in 1901 and quickly expended the bulk of its resources into draining the underground reservoir. The quantities of water removed from Tombstone were remarkable for the arid Southwest, exceeding one million gallons per day by 1909.[18] In the best of circumstances, pumping water out of mine shafts was a difficult and expensive proposition, but the DCA had enough money to purchase heavy-duty pumps, and enough men to run them and to address the technical problems that inevitably occurred at an industrial site.[19]

By financing the work of the Tombstone Consolidated Mining Company, in particular the pumping of water, the DCA enabled Tombstone to exist as a mining camp for several years beyond the time when a less diversified company would have had to pull out. In addition, the DCA brought into Tombstone large quantities of heavy industrial equipment. This equipment included four boilers weighing twenty-five tons apiece, which required a team of thirty-four horses to haul up the road from the nearby rail depot at Fairbanks; all-new steam pumps capable of pumping over 1,700 gallons of water per minute; and a nonflammable steel pump house, which enabled the Tombstone Consolidated to mine ore bodies below the water table without risking a fire. Mining engineer W. P. Blake noted in a report on the property prepared shortly after the DCA began setting up in Tombstone, "The great advantages resulting from the consolidation of interests... are evident."[20] By centralizing the administration of the mine-draining operation, the cost of staffing the boiler-and-pump system was dramatically reduced. Such economy of scale was possible because the Development Company of America absorbed the high start-up and operating costs. The DCA, with its diverse holdings, was further able to stave off the expense of running the mine at Tombstone, as profits at its other mine sites could offset the everyday expenditures there.

As general manager, mining engineer William Staunton was pleased to have centralized authority over the work at the Congress, the Silverbell, and Tombstone, but he was less enthused by the fact that his own work at each location was also subject to significant oversight. He was accustomed to having the last word, and the excellent relationship he had with the president of the Congress mine, E. B. Gage (also the first president of the Development Company) predisposed him to think relations with the DCA would be equally friendly.[21]

Staunton's interactions with Frank Murphy, however, were plagued by mutual recriminations and misunderstandings. Although Staunton was a well-connected and successful mining engineer, his position within the DCA was less comfortable than it appeared. The radical centralization of DCA management—in which one board of directors had oversight of several distinct mine sites—meant that it was easy for the corporation, in the person of Frank Murphy, to call Staunton to task when there were technical or operational problems, which undermined Staunton's engineering and managerial expertise. Murphy micromanaged Staunton, insisting that the engineer allocate his time at the various properties according to Murphy's wishes. In the early years of the DCA, when they had a fairly respectful relationship, Murphy cloaked his directions with qualifications. To a suggestion from Staunton that the workings at the Congress be slightly rearranged, for instance, Murphy responded,

> Situated just as we are, all things considered, I am inclined to feel as stated in my telegram that it is going to be necessary for you to give considerable personal attention to conditions at Congress.... While I realize that you have a first class organization there, I cannot help but feel it was so that you could spend a little more time on the property, that better results would be obtained,—although I may be entirely mistaken. I don't think you can blame me for feeling this way, having as I do implicite [sic] confidence in your ability to meet and overcome difficulties.[22]

The question Murphy put to Staunton—could he spend more time at the Congress?—is illogical on its face. It is reasonable to suppose that as technical director, Staunton hired "a first-class organization" to oversee engineering details at Congress precisely so that he could spend *less* time at each individual mining site. Since the major problem for the Development Company was the Tombstone Consolidated Mining Company, surely Murphy would want his chief engineering officer, in whom he had "implicit confidence," to allocate his time according to technical necessity. Despite Staunton's ostensible authority over the engineering details of the workings at the DCA's mining operations, Murphy always had his eye on the bottom line, quibbling with Staunton's every move. Staunton felt perpetually disrespected by Murphy's incessant telegrams. "[A]n expression of dissatisfaction with the way the work has been managed," Staunton complained about Murphy's telegraphic griping, "can only tend to discredit a manager in the eyes of his subordinates, who necessarily see them [the telegrams], and this is advantageous neither to the company nor the standing of the manager."[23] The DCA turned a profit because it was diversified, but diversification made the technical oversight of each project more complicated. While the large copper companies built integrated systems for the efficient extraction of copper, development companies such as the DCA were principally about the efficient extraction of profits. Staunton continually raised questions about the efficiency of the DCA's engineering projects, while Murphy was concerned with the efficient working of the DCA as a whole.

The work at Tombstone did not proceed smoothly, which undoubtedly made Murphy's de facto dismissal of Staunton's engineering expertise particularly galling to the engineer, who was, as he put it, "ambitious to join the charmed circle" of the elite mining engineers of Arizona—James Douglas, T. A. Ricketts, and John Greenway.[24] Yet those men oversaw integrated mining operations that focused on the engineering of copper; the technical focus of the DCA was never so clear, and that ambiguity put Staunton, as technical manager, in a difficult position.

Adding to the conflict, Staunton and Murphy were tripped up by continued miscommunication about the financial situation at the different mines, exacerbating an already tense relationship.[25] The telegraphic codes employed by all mining operators at the time did not help matters, as an incorrectly translated transmission could cause considerable distress. On one occasion, Murphy understood a Staunton telegraph that he was unable to estimate the ore body at Silverbell with enough exactitude to "stand verification by possibly *hypercritical* examining engineer" to read "by possibly *hypocritical* examining engineer" and had a fit, presuming an offense to the examining engineer he had hired as a consultant. In this instance, as in countless others, Staunton begged Murphy's forgiveness, stating that he "guess[ed he] was unfortunate in choice of words."[26]

Matters came to a head in 1910 in a conflict over the purchase of approximately $40,000 of new pumping equipment, the installation and maintenance of which would then cost an estimated $200,000. The new equipment was needed because an accident, coupled with the heavy demands placed on the pumps, had rendered five of the Tombstone Consolidated pumps unworkable. Staunton declared that the value of the mines at Tombstone did not "warrant one in advising such an expenditure," and, furthermore, "We have reached a time when we ought to stop putting money into Tombstone until we get into a position from other operations to afford to carry on the work." He suggested that Murphy call in one or two consulting mining engineers to inspect the Tombstone operations for an outside opinion on the mine's viability.[27] Murphy castigated Staunton for his lack of faith in the Tombstone mines, asserting that Staunton bore sole responsibility for the failure of the pumps. The mining engineer resigned his position in the DCA. "Further consideration of his [Murphy's] letters seemed to me to make it practically necessary," Staunton observed shortly afterwards. "I have never been able to discuss our affairs with him in the plain way that it seems to me should be done without raising a storm and having my motives misconstrued and I am tired of it."[28]

It is not remarkable that Staunton had such vitriolic conflict with Murphy, as mining engineers often clashed with overly engaged financiers and company officials. It is notable, however, that Murphy's dismissal of Staunton's opinions and concerns suggests that Staunton's technical expertise was of secondary importance to his success at the DCA. Despite the reliance of the DCA on the work of Staunton and other engineers, Murphy's vision, and therefore the company's, was not of an efficiently run mining operation but of a fully integrated system whose primary goal was to make money and whose secondary goal was to mine ore, an inversion of the general work of mining engineers. As Staunton's superior, Murphy was able to undermine Staunton's expertise at will and to determine what was worth implementing and what was simply "expert advice," which Murphy could ignore since, after all, it was only *advice*.

The Guggenheim Exploration Company

The Guggenheim Exploration Company, which operated throughout Mexico with occasional forays into mining sites in places such as Utah, Colorado, and Alaska, had a similar effect on rail development in Mexico, particularly in the north, as the DCA had in Arizona. However, it was a fundamentally different kind of development organization. The Guggenheim family at the turn of the twentieth century have sometimes been called "the Copper Kings." Although this appellation suggests a greater control over the world copper market than the Guggenheims in fact possessed, it indicates the family's influence over copper extraction and refining in the Americas. Meyer Guggenheim, the family patriarch, made his second fortune mining silver in Leadville, Colorado, in the 1860s (his first wildly successful business was importing coffee to the United States). With his sons—all seven were active in the family business—Meyer soon capitalized on this initial success in ore extraction, establishing a smelter in Pueblo, Colorado, managed by a company named, somewhat confusingly, Philadelphia Smelting and Refining. M. Guggenheim Sons intended to use this smelter not only for processing local silver ores but also for working the rich silver-lead ores being mined south of the border. This plan was derailed in 1890 by the passage of the McKinley Tariff, which taxed imported silver-lead ore in an attempt to stabilize the price of silver in the United States, essentially excluding Mexican ore from the American market.[29] Undaunted, M. Guggenheim Sons took the next logical step and at the suggestion of mining engineer Edward Newhouse built two smelters in Mexico with the full cooperation of the Díaz administration, thereby bypassing the provisions of the tariff. The first, in Monterrey, opened in 1892; the second, at Aguascalientes, in 1894.[30]

Between the cost of labor in Mexican smelting—approximately one-fifth of that in Colorado—and the concessions the Guggenheims received from the Mexican government in the form of duty-free importation of equipment and reduced taxes, these two smelters were extremely profitable from the start. They became even more so as they began treating the copper ore that was also present in northern Mexico in large quantities.[31] Including the Guggenheim refinery in Perth Amboy, New Jersey, which refined treated copper from Colorado and Mexico for sale to the emerging electrical industry, the family was responsible for smelting, transporting, and finishing a large percentage of the copper produced in the borderlands through the 1890s.

This dominance was recognized by other smelter operators in the United States, who united to organize the American Smelting and Refining Corporation (ASARCO) in 1901. Invited to join, the Guggenheim family declined, choosing instead to focus on their new exploration business, the Guggenheim Exploration Company (Guggenex). The family's reasons are not entirely clear. Possibly they believed that absent a major technological breakthrough in ore processing, current metallurgical practice was unable to satisfactorily process the lower-grade copper ore that was increasingly being mined through the southwestern Rocky Mountains and Mexico.[32] Possibly the family simply did not want to be involved in any business in which they did not own at least 51 percent. Regardless, without the participation of the Guggenheims, ASARCO folded, and the owners had to beg the Guggenheim family to buy them out—which they did. After 1907, the Guggenheims held not only Guggenex and their own smelter operations but also a controlling interest in ASARCO.[33]

The Guggenheim Exploration Company was capitalized in 1899 at $6 million "to prospect, explore, improve, and develop mining properties in any part of the world."[34] Under the leadership of the extremely energetic Daniel Guggenheim, Guggenex hired John Hays Hammond, a mining engineer already famous for his exploits in Africa with Cecil Rhodes, to oversee the surveying operations, paying him a base salary of over $200,000, which was reportedly the highest in the world at the time, in addition to significant shares in any mines the family decided to develop at his suggestion.[35] Guggenex was modeled after the London-based, Rothschild-financed Exploration Company, which had been exploiting mine sites principally in Latin America, Africa, and Australia since the late 1880s. Like the Development Company of America, the Exploration Company used its capital to finance site inspections, but rather than holding properties itself, it sought other investors and expertise to develop the most promising locations. More traditional mining companies were formed in the wake of significant finds to exploit particular veins and ore deposits. The Exploration Company sought to systematize the process of finding these potentially lucrative mining sites. Many of the mining engineers who were closely involved in the Exploration Company were American, and its work was well known in the United States.[36]

Daniel Guggenheim was widely believed to trust in the expertise of his technical employees and to rely on the advice of experts rather than on his own knowledge of the industry when making investments in mining—the opposite of Frank Murphy's relationship to William Staunton in Arizona.[37] It is certainly the case that the Guggenheims hired many of the most well-known and heavily credentialed mining engineers in the business and paid their technical experts—known in the business as "Guggies"—extremely generously. The high wages were possible due to the low overhead of running the Guggenheim Exploration Company, which paid engineers and their assistants but did not have the equipment and labor expenses of holding companies such as the DCA or mining companies such as Cananea Consolidated.[38]

For most of the nineteenth century, consulting engineers were hired directly by a board of directors or a set of investors to survey specific mine sites. Even before hiring the engineer, these investors were somewhat committed to the mine. After all, they were interested enough to hire an expert to scope out what was already in place with a view toward the future. Although the Guggenheim Exploration Company sometimes surveyed extant mine sites for specific clients, it usually worked on a completely different model. The firm decided where to send its engineers. These engineers worked for their manager and reported to him, rather than to a set of investors or an outside company. Managers were themselves experienced engineers who had many ideas as to where the best mines might be located.

They sent teams of surveyors and assayers to those locations and, if the results were promising, sought investors.

Guggenex thus upended the traditional expert-client relationship. The examining engineers they hired only had one employer, which established very specific requirements for what constituted a good prospect. Having such clear standards, in turn, should theoretically have eased the strain of writing mine reports, as the consultants knew exactly what information was required and how the report's recipient would interpret it. These engineers were no longer writing reports for investors or potential investors. Their reports would be read by a corporate supervisor with technical expertise. That supervisor, Alfred Chester Beatty, passed along information and recommendations, and it was up to Hammond and then Daniel Guggenheim to decide whether a mine was a worthwhile proposition. This bureaucratized and streamlined a decision that in other situations was negotiated directly between a consulting engineer and an investor or investors.

Logically, working directly for a manager such as engineer A. C. Beatty ought to have made the everyday workload of Guggenheim engineers easier. After all, there were no conflicts of interest in this relationship as there were between consulting mining engineers and an investment company, where the possibility of corruption—of the mining engineer simply finding the information that his employers wanted him to find—was high. Yet the pressure to find previously untapped mining resources, or clever new ways to mine extant-yet-undervalued mining deposits, was extremely high in Guggenex. Moreover, Beatty was not an easy man to work with. He had strong opinions about personnel deployment, and from a distance he sought to regulate very precisely how each engineer divided his time and used his team of assistants.[39] His directives could upend an established hierarchy by ordering an engineer to report to someone he considered his collaborator, or by dispatching a mining engineer's assistant to a mine a few hundred miles away, leaving the field operatives perplexed and infuriated.

Engineer Ross Hoffmann once received a telegram instructing him to forward immediately a report containing the assay results for a property in Chihuahua. Hoffmann told Beatty that it was with "great surprise... [I learned] that I am expected to report on any of the properties. I understood definitely that Mr. Gemmel was in charge of the work and I was under his orders." Being asked for a report he did not expect to have to prepare was of serious concern, for he was pressed for time and money. In this instance, Hoffmann had turned over his original and duplicate assays to Gemmel, a senior and very well-regarded mining engineer. The originals were sent to Monterrey to be assayed; the duplicates were in an unsealed sack in Gemmel's room, a circumstance Hoffmann thought might indicate that they had been tampered with. Hoffmann further complained to Beatty about rushing his work. He felt overextended as it was and, resentful of Gemmel's authority, was frustrated by almost every aspect of his survey work in Mexico.[40] Hoffmann noted that if he did not write his report in about half the time he thought he ought to take, his schedule would be derailed, and he and his fellow mining engineers would be unable to complete their designated work before the end of 1903.

When combined with the challenge of getting American dollars to Guggenex employees working at mine sites in remote locations, and the exhausting pace of mine inspections that Guggenex consultants maintained, it is clear that mining engineers frequently experienced the radical centralization of the Guggenheim exploring method as a mixed blessing.[41] Although their work was salaried, a luxury unknown to mining engineers who worked for less-solvent development companies, Guggenheim engineers had very little autonomy. Just three weeks prior to the telegram requesting Ross Hoffmann's assays, he drafted a letter of complaint to Beatty.

> I am acting now entirely under Gemmel's instructions according to your advice to him—in a recent letter[,] a portion of which was read to us.... As I am now in a position which requires very little responsibility...it will be a great relief and favor to me if you will send some-one to take my place—I feel that I need the rest ~~+ don't want to sacrifice my health under the circumstances~~ + ought not undertake this extended work under the circumstances.[42]

Hoffmann remained in Mexico for some time after writing this letter, and we can only speculate as to whether his complaints may have had the opposite result to that intended, or whether on further reflection he decided not to ask Beatty to move him to another assignment.

It is not at all surprising that Hoffmann was exhausted. Mine inspections were hard work, with or without bureaucratic frustrations. Mining engineers were expected to scrutinize the full extent of a property. In general, this meant that an engineer, along with his colleagues and assistants, walked or rode on horse or mule over the entire property. A well-planned inspection provided an accurate map of the correspondences between surface and underground points and elaborated with specific suggestions on how future work should proceed.[43] To achieve such a comprehensive survey of a mine, engineers had to find and record each outcropping, waterway, and elevation change, accurately depicting the latitude and longitude of neighboring claims in order to map the direction, depth, and breadth of the ore body. There were many means for determining the direction and extent of the ore body, as well as its composition. But when Guggenex was founded, the most common method was still to collect surface samples at regular intervals and send them to a chemist for assay. The largest mine survey undertaken by the Guggenheim Exploration Company, of a property belonging to the Utah Copper Company, required sixteen engineering assistants, took seven months, and cost $150,000, a staggering sum in 1903.[44]

The Technocratically Legible Landscape

The reports that the mining engineers of the Guggenheim Exploration Company submitted were extremely detailed, containing descriptions of mining properties and extant technology, analyses of local labor availability, and evaluations and estimates of future productivity. Many reports also contained exceptionally long introductions into both local culture and the history of a particular mining site. But the combination of economic and productivity analyses, personal letters to Beatty and each other that accompany these reports, and frequent progress updates gives us a comprehensive vision of the mining borderlands.

The landscape that emerges from engineers' reports of mining districts in Mexico and Arizona is not quite a terra incognita. Rather, it is a region that lacked vitality, as evidenced by their constant efforts to suggest new ways to increase productivity. There is a broad jingoism and ethnocentrism underlining observations of slack management in these reports that mirrors ethnic, class, and racial prejudices common among elite white Americans at the turn of the century.[45] A recurring trope in engineers' reports from Mexico, for instance, was that of a landscape gradually coming under the influence of a rationalizing force but one that periodically spun out of control. A mining district might have been worked by the Spanish and then by Mexicans "in a very unsystematic and unscientific manner" albeit with great "clever[ness] at following ore bodies…cheaply."[46] One copper mine was described by its inspecting engineer as having been "worked by the Mexicans in a desultory manner for the past 25 years."[47] The labor of Mexicans was thus trivialized or dismissed, although the fact of its existence served as an important marker of a mine's potential productivity. Engineers well understood that a mine that had produced significant ore when excavated by hand and treated using the patio process had the potential to produce considerably more when excavated with machinery and treated using the latest smelting technology, and they said as much in their reports.[48]

Of greater significance than the condescension of American engineers toward Spanish and Mexican mining methods was the quantification and abstraction of the landscape, the consulting engineer's primary methodological device. A 1905 report by a Guggenheim engineer on the Copper Chief Mine in Jerome, Arizona, for instance, consists of little more than a line sketch of a vertical section of the mine, coupled with a computation of the quantity of developed ore in the main ore body, plus the probable quantity of ore in an as-yet-undeveloped ore body, divided by an estimate of the cost of extraction, per ton. In this case, the estimated net value of the mine was $900,000.[49] Such quantification of mining terrain was the reason mining engineers were sent into Mexico. These were numbers that only such technically trained men could produce. The reports contained data points that were of tremendous utility, not only for the Guggenheims as they determined which mines to purchase but also for those mining engineers, Wall Street bankers, and reporters who watched the actions of the Guggenheims and their several businesses with interest, gleefully noting the incursions of American-backed finance into Mexico.[50]

James Scott has described the use of highly quantified and rationalized maps and grids to promote the "legibility" of landscapes and populations in the twentieth century as an attribute of "high modernism," a technocratic worldview he identifies as a precondition for control of a subject population. Scholars of engineering have argued that engineers embody the ideology of high modernism as described by Scott, possessing a "self-confidence about scientific and technical progress, the expansion of production, the growing satisfaction of human needs, the mastery of nature...and, above all, the rational design of social order commensurate with the scientific understanding of natural laws."[51] The Guggenheim mining engineers obviously conform to this standard: they were agents of rationalism, scanning the landscape to bring technical progress to places that by virtue of their inaccessibility had previously escaped industrialization. Mining, particularly on the scale undertaken by Guggenheim interests in the twentieth century, is ultimately and horrifically about human mastery over nature.[52] The ores of interest to the Guggenheims, principally silver and copper, are two metals that certainly contributed through their monetary value and significance to the new electrical industry to "the growing satisfaction of human needs."

Yet Guggenheim mining engineers lacked two key elements of the "high modernism" cited by Scott and others: they were not agents of a state; and although cognizant that a large mining project supported by the Guggenheims would lead to a reorganization of the local labor force and the likely relocation of many workers, the engineers of the Guggenheim Exploration Company were not interested in social reorganization per se. In his account of the Arabian American Oil Company (a subsidiary of Phelps Dodge), Robert Vitalis argues that the history of large corporate firms and the history of states are not dissimilar. Discussing the racialized division of labor common in the mining industry at the turn of the twentieth century, Vitalis notes that large corporations in the southwestern United States, such as Phelps Dodge, "organized production in the way that the post-Reconstruction South organized society."[53]

As an exploration and development company, Guggenex was not as deeply enmeshed in building racially stratified mining enclaves as Phelps Dodge or the Arizona Copper Company, but company engineers were practiced at cataloguing information that enabled such stratification: population; proximity to towns and major roads; access to resources such as water and timber. Guggenheim engineers performed a job similar to that of the Corps of Topographical Engineers, dispatched westward to survey new American territory in the wake of war with Mexico, or to John Russell Bartlett's survey of the U.S.–Mexico border, during which the cataloguing of mineral resources of the new territory was almost as important as the survey of the international boundary.[54] Guggenheim mining engineers produced a quantified and particularist reading of the mining landscape that clarified logistical and technical needs and enabled future exploitation.

The scale of the Guggenheim Exploration Company was extraordinary, easily dwarfing the contemporary work done by more traditional mining companies. In 1910, Guggenheim engineers conducted over 1,600 mine inspections, including preliminary examinations of 268 mines and full surveys of seventy-four mines.[55] In that year, the Guggenheim Exploration Company had been in operation for a decade, and the accretion of information on the landscape of the borderlands mining districts during this time was phenomenal. A small army of engineers and assistants was required to work at a frantic pace to accomplish all these surveys. No wonder Ross Hoffmann worried that he would fall behind in his inspections. As comprehensive as was the information gathered through mine inspections and surveys, Guggenheim mining engineers paid surprisingly little attention to local or regional politics. Indeed, this lack of attention was fairly anomalous in the mining industry. It is impossible to believe that businessmen as practiced and knowledgeable as the Guggenheims did not consider the legal and political implications of every given investment. U.S. tax laws and concessions granted to foreign investment were important reasons that the Guggenheims invested in their first smelter in Mexico. Engineering impresario John Hays Hammond counted Porfirio Díaz among his personal friends, and this relationship smoothed the way for many Guggenheim ventures in Mexico.[56] Yet it is also notable that considerations of politics and legislation seem to have been beyond the purview of their workaday consulting engineers. In fact, the narrowness of the "Guggies'" consideration of mining sites almost amounts to a willful effort on the part of individual engineers to ignore the effects of the vast infusion of finance portended by the interest of the Guggenheim Exploration Company.

As the borderlands mining industry matured and expanded in the early years of the twentieth century, companies such as DCA and Guggenex bureaucratized the work of mining engineers toward functions as managers and consultants on a much larger scale than was conceivable in the 1870s and 1880s. Although these companies were utterly reliant on the expertise and professionalism of mining engineers, they narrowed the workload of individual engineers, much as did the consolidated mass-mining companies. At the same time, development and exploration companies helped to produce a legible technocratic landscape that served as a blueprint for future exploitation, thus providing work for future generations of mining engineers.

Notes

1. Robert Vicat Turrell with Jean-Jacques Van Helten, "The Rothschilds, the Exploration Company, and Mining Finance," *Business History* 28, no. 2 (1986):181–183.
2. LaFeber, *The New Empire*, 57–59.
3. Paul Sabin, "Home and Abroad: The Two 'Wests' of Twentieth-Century United States History," *Pacific Historical Review* 66, no. 3 (1997):314–315; 318.
4. Veblen, *Engineers and the Price System*, 74.
5. Veblen, *Engineers and the Price System*, ch. 2; Edwin Layton, "Veblen and the Engineers," *American Quarterly* 14, no. 1 (1962):64–72; Layton, *Revolt of the Engineers*, 225–226.
6. J. M. Seibert to Knauss, May 17, 1901, J. M. Seibert Letterbook 1899–1902, Mss. HM 63953, HL.
7. Seibert to Dan Turner, September 13, 1901, J. M. Seibert Letterbook.
8. J. M. Seibert to J. V. Keeley, March 8, 1902, Seibert Letterbook.
9. J. M. Seibert to George B. Clark, March 31, 1902, Seibert Letterbook.
10. The archival trail on Seibert's company runs out with Seibert's letterbooks, which are held at The Huntington Library. It is possible that the company had some modest success, but as with most mining ventures, it never fulfilled the investors' golden dreams.
11. "List of Recent Graduates," *Columbia School of Mines Quarterly* 5, no. 2 (January 1884):162.

12. William P. Blake, *Tombstone and its Mines* (New York: Development Co. of America, 1903), 15–20; Staunton, "Memoir," WFS.

13. Robert L. Spude, "Frank Morrill Murphy, 1854–1917: Mining and Railroad Mogul and Developer of the American Southwest," in *Mining Tycoons in the Age of Empire: Entrepreneurship, High Finance, Politics, and Territorial Expansion*, ed. Raymond E. Dumett (Burlington, VT: Ashgate, 2009),156.

14. Spude, "Frank Morrill Murphy," 152–153, 157–159.

15. In size, the DCA was roughly equivalent to that of other mining companies in the region, including Greene-Cananea and the Guggenheim Exploration Company, although it remained somewhat smaller than Phelps Dodge. Until the 1920s, all the southwestern and Mexican companies were small compared to Anaconda, the company that controlled the mines in Butte, Montana, and had assets of over $170 million in the early twentieth century. Norman R. Collins and Lee E. Preston, "The Size Structure of Large Economic Firms," *American Economic Review* 51, no. 5 (1961):1005; 1008.

16. Staunton, "Memoir," 105, 107–109; V. L. Mason to William F. Staunton, November 11, 1905, folder 6, box 3, both WFS; Spude, "Frank Morrill Murphy," 156–157.

17. William F. Staunton, "Report, Tombstone Mill and Mining Co." (1888); William F. Staunton, "Report Tombstone Mill and Mining Co." (1889); William A. Farish, "Report on the Mines of the Tombstone Consolidated Mines Co., Ltd." (1901).

18. William F. Staunton to V. L. Mason, January 31, 1909, folder 6, box 3, WFS.

19. William F. Staunton to Development Company of America, June 16, 1903, folder 6, box 3, WFS.

20. Blake, *Tombstone*, 19–20.

21. [Memoir], 73, 77–80, 88.

22. Frank Murphy to William F. Staunton, January 9, 1904, folder 1, box 3, WFS.

23. William F. Staunton to Frank Murphy, May 5, 1910, folder 5, box 3, WFS, UASC.

24. Staunton, "Memoir," 177, WFS, UASC.

25. William F. Staunton to Henry Robinson, December 26, 1908, folder 1, box 4, WFS.

26. William F. Staunton to Frank Murphy, December 5, 1904, folder 1, box 3, WFS.

27. William F. Staunton to Frank Murphy, May 8 and May 21, 1910, folder 5, box 3, WFS.

28. William F. Staunton to Henry Robinson, May 17 and May 23, 1910, folder 2, box 4, WFS.

29. Irwin Unger and Debi Unger, *The Guggenheims: A Family History* (New York: Harper, 2005), 39–40.

30. Unger and Unger, *The Guggenheims*, 41–42; Isaac F. Marcosson, *Metal Magic: The Story of the American Smelting and Refining Company* (NY: Farrar, Strauss, 1949), 45–51.

31. Thomas O'Brien, "Copper Kings of the Americas—The Guggenheim Brothers," in Dumett, ed., *Mining Tycoons*, 203.

32. O'Brien, "Copper Kings," 211. This seems unlikely, since in the early 1900s the Guggenheim family had a stake in the Bingham Canyon Mine in Utah, a clear example of significant technological development in processing low-grade ore, as mining engineer Daniel Jackling was rapidly transforming it into an open-pit copper mine.

33. Marcosson, *Metal Magic*, 61, 63, 69.

34. Marcosson, *Metal Magic*, 63 [Notice of Incorporation of Guggenheim Exploration Company]; *Engineering and Mining Journal* 67, no. 23 (June 10, 1899):686.

35. John Phillips, "Alfred Chester Beatty: Mining Engineer, Financier, and Entrepreneur, 1898–1950," in Dumett, ed., *Mining Tycoons*, 220. On Hammond in Africa, see Jessica Teisch, *Engineering Nature: Water, Development, and the Global Spread of American Environmental Expertise* (Raleigh: University of North Carolina Press), 97–131.

36. Spence, *Mining Engineers*, 137–138; Teisch, *Engineering Nature*, 98–100.

37. Isaac Marcosson first made this point in his admiring history of ASARCO, but it was picked up by later biographers and hagiographers of the Guggenheim family. It falls short in explaining the public and extremely acrimonious split between Guggenex and mining engineer A. C. Beatty, who accused Daniel Guggenheim of inflating the stock price of a prospect in Alaska on the basis of what Beatty believed to be a preliminary and less-than-enthusiastic report. Marcosson, *Metal Magic*, 45, 46; Anon., *The Men Who Are Building America*, reprint (New York: Read Books Design, 2009), 177; see also Christopher Schmitz, who argues that technological innovation, such as that supported at Guggenheim-owned facilities, rather than financial innovation, was the best indicator of success in the copper smelting industry in the early twentieth century. Schmitz, "The Rise of Big Business in the World Copper Industry, 1870–1930," *Economic History Review*, new series 39, no. 3 (1986):392; 404.

38. Charles Harvey and Jon Press, "Overseas Investment and the Professional Advance of British Metal Mining Engineers, 1851–1914," *Economic History Review*, new series 42, no. 1 (1989):74; Spence, *Mining Engineers*, 138.

39. A. C. Beatty to Ross Hoffmann, January 15, 1903, folder 32, box 2, MS 3163, AHC.

40. Ross Hoffmann to A. C. Beatty, July 1, 1903; Longacre to Ross Hoffmann, July 4, 1903; [Robert C.] Gemmel to Ross Hoffmann, July 8 and 19, 1903, all folder 32, box 2, MS 3163, AHC (hereafter RH).

41. Ross Hoffmann to A. C. Beatty, July 1, 1903, folder 32, box 2, RH.

42. Ross Hoffmann to A. C. Beatty, June 6, 1903, folder 32, box 2, RH.

43. Charles F. Hoffmann notebook, [n.d.], folder 10, box 1, RH; Arthur Laing to James Douglas, July 18, 1881, folder 29, box 3, JD; James D. Hague to Santa Eulalia Mining Co., May 20, 1885, L-9, JDH.

44. Spence, *Mining Engineers*, 91, 96–99. Nystrom, *Seeing Underground*, 9–10.

45. Reginald Horsman, *Race and Manifest Destiny: The Origins of American Racial Anglo-Saxonism* (Cambridge, MA: Harvard University Press, 1981), 210; Matthew Frye Jacobsen, *Whiteness of a Different Color: European Immigrants and the Alchemy of Race* (Cambridge, MA: Harvard University Press, 1998), 90–95; Deutsch, *No Separate Refuge*, 110–111.

46. [R. C. Gemmel], "Preliminary Report...," December 24, 1901, folder 4, box 6, Harold A. Titcomb Papers, MS 2220, American Heritage Center (hereafter HAT).

47. John M'Intyre to S. D. Bridge, August 19, 1899, folder 13, box 6, HAT.

48. John M'Intyre to S. D. Bridge, August 19, 1899, folder 13, box 6, HAT; [unknown], "Preliminary Report on the Smelter at Bonanza, Zac...," December 24, 1901, folder 4, box 6, HAT; C. W. Geddes, "Report on a Reconnaissance of the Mammoth Mine," folder 19, box 6, HAT; W. A. Farish, "Report on the Mina Grande and Santa Teresa Properties in the District of Hermosillo..." June 23, 1902, folder 20, box 6, HAT; W. A. Farish, "Report on the Zaragosa Properties...," July 7, 1902, folder 17, box 7, HAT.

49. [Harold A. Titcomb], Copper Chief Mine Report, May 20, 1905, folder 4, box 2, HAT.

50. "Millions for Mexico," *New York Times*, April 1, 1901; "Millions for Mexican Mines," *New York Times*, November 1, 1902; "Guggenheim's Roads," *Los Angeles Times*, January 23, 1903; "Guggenheim Exploration Company Increases its Capitalization," *Los Angeles Times*, March 6, 1903.

51. James Scott, *Seeing Like a State: How Certain Schemes to Improve the Human Condition Have Failed* (New Haven, CT: Yale University Press, 1999), 4; Martin Reuss, "Seeing Like an Engineer: Water Projects and the Mediation of the Incommensurable," *Technology and Culture* 49, no. 3 (2008):543.

52. LeCain, *Mass Destruction*.

53. Robert Vitalis, *America's Kingdom: Mythmaking on the Saudi Oil Frontier* (Stanford: Stanford University Press, 2007), 19.

54. Paternalistic policies at Phelps Dodge, for instance, led to the creation of a stratified racial social sphere and to the labeling of Bisbee as a "white man's camp." There is no evidence indicating that these social policies were undertaken for specific technical purposes. Indeed, the fact that Phelps Dodge's technical work was emulated all over the world, while the racial policies at Bisbee only occurred in locations at which management had a particularly paternalistic bent, suggests that mining engineers, as a *profession*, were not particularly vested in paternalism. See Benton-Cohen, *Borderline Americans*, on Bisbee, and Truett, *Fugitive Landscapes*, ch. 6, on the landscape architecture of Cananea, for detailed discussion of how race affected the built environment in borderlands mining towns. William Goetzmann, *Army Exploration and the American West, 1803–1863* (New Haven, CT: Yale University Press, 1959), 1, 4, 10; John Russell Bartlett, *Personal Narrative of Explorations and Incidents in Texas, New Mexico, California, Sonora, and Chihuahua, Connected with the United States and Mexican Boundary Commission during the Years 1850, '51, '52, and '53* (New York: D. Appleton, 1854), iv–v; Rebert, *La Gran Linea*.

55. Spence, *Mining Engineers*, 138.

56. Mark Wasserman, *Capitalists, Caciques, and Revolution: The Native Elite and Foreign Enterprise in Chihuahua, 1854–1911* (Chapel Hill: University of North Carolina Press, 1984), 86–89.

CONCLUSION

Mediating Resources

As Commissioner for the United States and Mexico boundary survey in the early 1850s, John Russell Bartlett took it upon himself to make a study of the natural resources of the deserts and mountains in the territory so recently acquired by the United States. "The time is not far distant," he observed, "when crowds as large as those now pressing on to California and Australia will be 'prospecting' among the mountains of Texas, New Mexico, Chihuahua, and Sonora, attracted by similarly rich deposits, and probably with the like splendid success."[1] When American entrepreneurs took up Bartlett's implicit challenge in the southwestern borderlands, they certainly brought with them dreams of wealth shaped by tales of the California gold fields. And they did find rich mining opportunities, but ones that would require significant work to make them pay. For the next sixty years, mining development appeared to follow a progressive trajectory: mining prospects replete with refractory, low-grade ores were exploited using ever-more sophisticated equipment and processes. The active work of mining engineers as technical experts who mediated capital investments in the region, however, was crucial to actualizing Bartlett's call. In turn, the success of this project led to a transformation of the work of mine engineering.

The development of the borderlands mines did not follow the pattern of the California gold rush. In 1864, when journalist J. Ross Browne traveled across southern Arizona, he had quite an adventure, but he was disappointed in the mining opportunities he found.

> The great drawback to mining [here] is, that the owners of feet [mining claims] have no money to expend in extracting their wealth from the ground; and when people who have money desire to invest, the men of feet demand extraordinary sums, because they think claims that attract capital must be of extraordinary value.[2]

In Browne's assessment, the borderland mines lacked owners who were knowledgeable about finance or had regular access to capital; therefore little development occurred. His observation was consistent with what mining engineers such as Raphael Pumpelly and independent prospectors such as John Denton Hall also found there in the late 1850s and 1860s. Thus, through the 1870s and 1880s, with the exception of a few booms, such as the silver strike at Tombstone, only companies with deep pockets were more than marginally successful at working the precious metals of Arizona and Sonora, while the independent "men of feet" continued to prospect around, occasionally getting lucky. This was in contrast to some of the other, more established mineral regions of the country, such as Butte or Leadville, where ample investment funds enabled interested financiers on the ground to import "experts," either for mining or litigation, depending on the situation.

The advent of electricity in the 1890s, however, dramatically changed the market for copper ore and fundamentally altered the economics of the region. Copper mining required a massive up-front investment, and the financially naïve mine owners described by Browne were bought out by well-connected investors and corporations from New York and San Francisco. Crucially, it was mining engineers who made the borderlands legible to investors by means of their newly scientific underground surveys; persuaded investors of the critical importance of building infrastructure to aid in mineral extraction; and brought heavy technology to mining sites, building a new industrial landscape.

Mining engineers did not merely change and exploit the landscape, however. They proved more than willing to bend their careers to fit the needs of their new hierarchically organized, technocratic patrons. From the time the earliest mining engineers traveled to the Santa Rita, Mowry, and Heintzelman Mines at Tubac, through the Mexican Revolution, the career trajectory of mining engineers changed dramatically. Early engineers, members of an elite and cosmopolitan class of East Coast and European men, traveled extensively through their careers, working as consultants or managers as it suited them or as the opportunity arose. For these engineers, performing an individualistic, western, rugged, masculine identity was crucial to their work as technical experts. However unlikely the adoption of a self-consciously pioneering masculine identity might have been for a profession of mostly elite white men from the eastern United States, it stands to reason that they would work to find ways to buttress their authority in the field.

Despite the undoubted mineral wealth of North America, the profession of mine engineering was slow to take hold in the university system, relative to its sister fields of civil or mechanical engineering. Perhaps because of this, the position of mining engineers as "experts" was tenuous through the late nineteenth century. Whether conducting a mine survey or working as the resident mine manager, mining engineers had to position themselves as authoritative technical workers vis-à-vis working miners when in the field. Mere credentialing held little sway when the workers were experienced Cornish miners, and was even harder to assert to a population of immigrant miners who did not speak English. Embracing an identity as westering pioneers enabled mining engineers to assert themselves with a distinctive regional authority. When interacting with eastern capitalists as well, mining engineers found that they established confidence if they could make it clear that they were equally at home in rugged mining camps and the social clubs of Manhattan or San Francisco.

By the 1910s, mining engineers in the borderlands were more likely to work for the large copper corporations such as Arizona Copper, Phelps Dodge, and Cananea Consolidated Copper than they were to manage business on site for smaller companies. In these new, heavily industrialized mines, engineers worked in teams as surveyors and technicians, developing ever-more efficient machinery and means of extraction. Although the more independent aspects of the profession did not disappear, fewer mining engineers were called upon to utilize the breadth of their skill set. More often, they spent their careers as corporate functionaries, interacting with other engineers and overseeing mine labor and operations. Even the engineers who remained most mobile, such as those employed by the Guggenheim Exploration Company, worked as part of an army of experts rather than as individuals forced to assert their own credentials and professional worth.

The complex identity work undertaken by nineteenth-century mining engineers was thus no longer as important except as a recruiting tool for new members of the profession, since the romance of the Old West still proved enticing. With the industrialization of mining through the borderlands and the corporatization of mining companies, mining engineers no longer had to prove they were worth the expense. Rather, they were fully integrated into the corporation in an industry that could not have been established without their initial translational work. The success of a cadre of elite mining engineers in constructing technologically sophisticated mining systems that rendered mines legible to large capital investors had the effect of devaluing "the local" in their work and in the landscape, narrowing the scope of the profession for generations of mining engineers who followed them, and fundamentally altering the meaning and appearance of the borderlands.

When mining engineers headed out to Arizona or Sonora in the mid-to-late nineteenth century, they found a place and a society that had little in common with the northeastern United States and Europe. Prior to the 1880s, neither the U.S. nor the Mexican governments was able to adequately protect their citizens from Apache raiding parties, which quite understandably viewed the encampments of foreigners as incursions into their sovereign territory. As Raphael Pumpelly's memoir demonstrates, the high adventures to be found working in such circumstances were not always comfortable, and the political and social circumstances of the early borderlands did not facilitate great advances in the mining business or in mining techniques.

Yet by the 1920s, largely as a result of the industrial push of copper mining, engineer Ralph Ingersoll was able to travel by train to his position in Pilares de Nacozari in great comfort, particularly once he crossed into Mexico where he was permitted to smoke on the train and put his feet up on the seat in front of him.[3] At Nacozari, he found a robust community of fifty Americans, including two female schoolteachers and some extremely enthusiastic golfers. This is a tremendous contrast to the experience of Morris Parker, merely twenty years earlier, who suffered, with his wife and three small children, through an uncomfortable, multiday trip involving a couple of different trains and a difficult wagon journey to reach Nacozari. Parker left the position in less than eighteen months because there were only five Americans in the camp and nobody with whom, in his estimation, his wife could carry on a conversation. In the early years of the 1900s, mining engineer George Kingdon and his family lived for years in full middle-class comfort in Cananea, Sonora, in a white house with a lawn and a Chinese cook, before his wife and children fled north to escape the violence of the Mexican Revolution.[4] As privileged members of the Anglo middle class, the lifestyles enjoyed (or not) by mining engineers at mining camps through the years is a superb signifier of the extent to which the region had been reshaped in the image of the metropolitan United States.

The impressive remnants of the major mining corporations of the early twentieth century such as Phelps Dodge, Cananea Consolidated, and Arizona Copper are self-evident in the borderlands today: in crumbling industrial buildings, polluted rivers, and the monstrous open pits and tailings piles that dot the terrain. The lasting effects extend beyond the physical, in the prevalence of place names that refer to the mining industry and corporations that dominated the economy from the late-nineteenth century onwards.[5] Such industrial detritus results from the mediating work of mining engineers in crafting the technocratic landscape that enabled an explosive capitalist expansion in the borderlands.

As industrial operators in the U.S. Southwest and into Mexico, mining engineers were successful innovators, devising new and ingenious ways to extract mineral wealth from a region of low extant ore, utilizing newly codified, and accessible, engineering knowledge. This engendered a change in mining. From an activity that required skilled workers, it became one dominated by nonselective, industrial operations overseen by bureaucratic, professional mining engineers lodged in corporate hierarchy. Thus, mining engineers were the critical mediators who enabled the growth, bureaucratization, and corporate consolidation of the borderlands mining industry, transforming the region into a technocratic landscape legible to capital interests.

Notes

1. John Russell Bartlett, *Personal Narrative of Explorations and Incidents in Texas, New Mexico, California, Sonora, and Chihuahua, Connected with the United States and Mexican Boundary Commission During the Years 1850, '51, '52, and '53* (New York: D. Appleton, 1854), iv–v.
2. Browne, *Adventures in the Apache Country* (New York: Harper and Bros., 1871), 72.
3. Ingersoll, *In and Under Mexico*, 7–15.
4. Maud Kenyon Kingdon, *From Out of Dark Shadows* (San Diego: Frye and Smith), 72–73.
5. Richard V. Francaviglia, *Hard Places: Reading the Landscape of America's Historic Mining Districts* (Iowa City: University of Iowa Press, 1991), 24–27; Robert G. Varady, Helen Ingram, Lenard Milich, "The Sonoran Pimería Alta: Shared Environmental Problems and Challenges," *Journal of the Southwest* 37, no. 1 (1995):120; John Harner, "Place Identity and Copper Mining in Sonora, Mexico," *Annals of American Geographers* 91, no. 4 (2001):660–680.

BIBLIOGRAPHY

Manuscripts and Special Collections

American Heritage Center, Laramie, Wyoming
- Frank A. Ayer papers (3341)
- Ross Hoffmann papers (3163)
- Henry D. Smith papers (7721)
- Harold A. Titcomb papers (2220)

Arizona Historical Society, Tucson, Arizona
- Cananea Consolidated Copper Company records (MS 1032)
- Cananea Mining Papers (MS 1311)
- Courtenay DeKalb papers (MS 1176)
- James A. Douglas papers
- John and Isabella Greenway collection (MS 311)
- Joseph Obermuller Collection (MS 593)
- Eugene Sawyer papers (MS 360)

Beineke Library, Yale University, New Haven, Connecticut
- Jarvis Robinson family papers (addition)

Harvard University Archives, Cambridge, Massachusetts

The Huntington Library, San Marino, California
- Henry D. Bacon papers
- John Daniell papers
- Mary Halleck Foote papers
- James D. Hague papers
- Louis Janin papers
- Louis Janin papers (addenda)
- Morris B. Parker papers
- Phelps Dodge Corporation collection of research material
- Raphael Pumpelly papers
- John R. Robinson Diary (HM 62476)

New-York Historical Society, New York, New York
- Eben Erskine Olcott papers (BV Olcott)

C. L. Sonnichsen Special Collections, University of Texas El Paso, El Paso, Texas
- Oral History Collection
- Hyslop-Beckmann family papers (MS 211)

University of Arizona Special Collections, Tucson, Arizona

William F. Staunton papers (AZ 152)
Lewis Douglas Papers (AZ 290)
Yale University Library, New Haven, Connecticut
Records of the Sheffield Scientific School (RU 819)

Newspapers and Periodicals

Arizona Daily Star [Tucson]
Arizona Miner [Prescott]
Bulletin of the American Institute of Mining Engineers
Chicago Tribune
Colorado School of Mines Magazine
Engineering and Mining Journal
Congressional Globe
Harper's Weekly
Mining and Scientific Press
Mining Magazine
Pacific Coast Annual Mining Review and Stock Ledger
Putnam's Magazine
New-York Daily Times
New York Sun
New York Times
New Yorker
San Francisco Sentinel
School of Mines Quarterly
Transactions [American Institute of Mining Engineers]
Weekly Arizonian [Tubac]
Weekly Arizonian [Tucson]

Books, Pamphlets, and Articles

[anon.] *Gold and Silver Mining in Sonora, Mexico: Proposed Purchase of the San Juan del Rio Mines and Lands, Belonging to The Cincinnati and Sonora Mining Association.* Cincinnati: Wrightson, 1867.

———. *Prospectus of the Loma de Platta, or Hill of Silver Mine, of Mexico, Situated in Altar District, State of Sonora, Republic of Mexico.* Atlantic City, NJ: Telegraph Steam Printing House, 1880.

———. *The Men Who Are Building America.* Reprint. New York: Read Books Design, 2009.

Acuña, Rudolph F. "Ignacio Pesquiera: Sonoran Caudillo." *Arizona and the West* 12, no. 2 (1970):137–172.

Andrews, Thomas. *Killing for Coal: America's Deadliest Labor War.* Cambridge, MA: Harvard University Press, 2010.

Ash, Eric. *Power, Knowledge, and Expertise in Elizabethan England.* Baltimore, MD: Johns Hopkins University Press, 2004.

———. "Introduction: Expertise and the Early Modern State," *Osiris*, 2nd series, no. 25 (2010):1–24.

Bakken, Gordon. *The Mining Law of 1872: Past, Politics, and Prospects.* Albuquerque: University of New Mexico Press, 2008.

Balch, William Randolph. *Mines, Miners, and Mining Interests of the United States in 1882*. Philadelphia: Mining Industrial Publishing Bureau, 1882.

Bancroft, Hubert Howe. *History of Arizona and New Mexico, 1530–1888*. San Francisco: A. L. Bancroft, 1889.

Basso, Matthew, Laura McCall, and Dee Garceau. *Across the Great Divide: Cultures of Manhood in the American West*. New York: Routledge, 2001.

Bartlett, John Russell. *Personal Narrative of Explorations and Incidents in Texas, New Mexico, California, Sonora, and Chihuahua, Connected with the United States and Mexican Boundary Commission During the Years 1850, '51, '52, and '53*. New York: D. Appleton, 1854.

Bederman, Gail. *Manliness and Civilization: A Cultural History of Gender and Race in the United States, 1880–1917*. Chicago: University of Chicago Press, 1995.

Berman, David R. *Radicalism in the Mountain West, 1890–1920: Socialists, Populists, Miners, and Wobblies*. Boulder: University Press of Colorado, 2007.

Benton-Cohen, Katherine. *Borderline Americans: Racial Division and Labor War in the Arizona Borderlands*. Cambridge, MA: Harvard University Press, 2009.

Bernstein, Marvin D. *The Mexican Mining Industry, 1890–1950*. Albany: SUNY Press, 1967.

Bijker, Wiebe, Thomas Hughes, and Trevor Pinch, eds. *The Social Construction of Technological Systems: New Directions in the Sociology and History of Technology*. Cambridge, MA: MIT Press, 1987.

Bimson, Walter R. *Louis D. Ricketts (1859–1940): Mining Engineer, Geologist, Banker, Industrialist, and Builder of Arizona*. New York: Newcomen Society of England, 1949.

Blackhawk, Ned. *Violence Over the Land: Indians and Empires in the Early American West*. Cambridge, MA: Harvard University Press, 2008.

Blake, William P. *Tombstone and Its Mines*. New York: Development Co. of America, 1903.

Bodnar, John. *The Transplanted: A History of Immigrants in Urban America*. Bloomington: Indiana University Press, 1985.

Brown, Jonathan C. "Foreign and Native-Born Workers in Porfirian Mexico." *American Historical Review* 98, no. 3 (1993):786–818.

Brown, Ronald. *Hard Rock Miners: The Intermountain West, 1860–1920*. College Station: Texas A&M University Press, 1979.

Browne, J. Ross. *Adventures in the Apache Country: A Tour through Arizona and Sonora, with Notes on the Silver Regions of Nevada*. New York: Harper and Bros., 1871. Reprint, New York: Promontory Press, 1974.

Buder, Stanley. *Capitalizing on Change: A Social History of American Business*. Chapel Hill: University of North Carolina Press, 2009.

Byrkit, James. *Forging the Copper Collar: Arizona's Labor-Management War of 1901–1921*. Tucson: University of Arizona Press, 1982.

Carnes, Mark C., and Clyde Griffen, eds. *Meanings for Manhood: Constructions of Masculinity in Victorian America*. Chicago: University of Chicago Press, 1990.

Chandler, Alfred. *The Visible Hand: The Managerial Revolution in American Business*. Cambridge, MA: Belknap Press, 1977.

Cleland, Robert Glass. *A History of Phelps Dodge, 1834–1950*. New York: Knopf, 1952.

Collins, Joseph Henry. *Principles of Metal Mining*. New York: G. P. Putnam's Sons, 1874.

Collins, Norman R., and Lee E. Preston. "The Size Structure of Large Economic Firms." *American Economic Review* 51, no. 5 (1961):986–1011.

Colquhoun, James. *The Story of the Birth of the Porphyry Coppers*. London: William Clowes and Sons, [1933].

Crampton, Frank. *Deep Enough: A Working Stiff in the Western Mine Camps*. Denver: Sage Books, 1956.

Cronon, William. *Changes in the Land: Indians, Colonists, and the Ecology of New England*. New York: Hill and Wang, 1983.

———. *Nature's Metropolis: Chicago and the Great West*. New York: Norton, 1992.

Curtis, Kent. *Gambling on Ore: The Nature of Metal Mining in the United States, 1860–1910*. Boulder: University Press of Colorado, 2013.

Day, David T. *Mineral Resources of the United States* [1902]. Washington, D.C.: Government Printing Office, 1904.

Deutsch, Sarah. *No Separate Refuge: Culture, Class, and Gender on an Anglo-Hispanic Frontier in the American Southwest, 1880–1940*. New York: Oxford University Press, 1987.

Doing, Park. *Velvet Revolution at the Synchrotron: Biology, Physics, and Change in Science*. Cambridge, MA: MIT Press, 2009.

Dumett, Raymond E., ed. *Mining Tycoons in the Age of Empire: Entrepreneurship, High Finance, Politics, and Territorial Expansion*. Burlington, VT: Ashgate, 2009.

Edwards, Rebecca. *New Spirits: Americans in the Gilded Age, 1865–1905*. New York: Oxford University Press, 2005.

Foster, J. W., and J. D. Whitney. *Report on the Geology and Topography of a Portion of the Lake Superior Land District in the State of Michigan*. Part 1, *Copper Lands*. Washington, D.C.: House of Representatives, 1850.

Francaviglia, Richard V. *Hard Places: Reading the Landscape of America's Historic Mining Districts*. Iowa City: University of Iowa Press, 1991.

Frehner, Brian. *Finding Oil: The Nature of Petroleum Geology*. Lincoln: University of Nebraska Press, 2011.

Fuller, John Douglas Pitts. *The Movement for the Acquisition of All Mexico, 1846–1848*. Baltimore, MD: Johns Hopkins University Press, 1936.

Garner, John S. ed. *The Company Town: Architecture and Society in the Early Industrial Age*. New York: Oxford University Press, 1992.

Geiger, Roger. *To Advance Knowledge: The Growth of American Research Universities*. New York: Oxford University Press, 1986.

Goetzmann, William. *Army Exploration and the American West, 1803–1863*. New Haven, CT: Yale University Press, 1959.

Gomez-Quiñones, Juan. *Mexican American Labor, 1790–1990*. Albuquerque: University of New Mexico Press, 1994.

Gordon, Linda. *The Great Arizona Orphan Abduction*. Cambridge, MA: Harvard University Press, 1999.

Greenberg, Amy. *Manifest Manhood and the Antebellum American Empire*. New York: Cambridge University Press, 1995.

Greeley, Michael N., and J. Michael Canty, eds. *History of Mining in Arizona*, vol. 1. Tucson: Mining Club of the Southwest Foundation, 1987.

Griffin, Larry J., Michael E. Wallace, and Beth A. Rubin. "Capitalist Resistance to the Organization of Labor Before the New Deal: Why? How? Success?" *American Sociological Review* 51, no. 2 (1986):147–167.

Grossman, Sarah E. M. *Capital Mediators: Mining Engineers in the Southwestern U.S. and Northern Mexico, 1850–1910*. PhD diss., University of New Mexico, 2012.

———. "Mining Engineers and Fraud: The U.S.-Mexico Borderlands, 1860–1910." *Technology and Culture* 55, no. 4 (2014):821–849.

Gunnell, John G. "The Technocratic Image and the Theory of Technocracy," *Technology and Culture* 23, no. 3 (1982):392–416.

Gutierrez, David. *Walls and Mirrors: Mexican Americans, Mexican Immigrants, and the Politics of Ethnicity*. Berkeley: University of California Press, 1995.

Hall, John Denton. *Travels and Adventures in Sonora: Containing a Description of Its Mining and Agricultural Resources, and Narrative of a Residence of Fifteen Years*. Chicago: J. M. W. Jones Stationery and Printing Company, 1881.

Hammond, John Hays. *The Autobiography of John Hays Hammond*, vol. 1. New York: Farrar and Rinehart, 1935.

———. *The Engineer*. New York: Scribner's, 1921.

Hancock, H. Irving. *The Young Engineers in Arizona; or Laying Tracks in the Man-Killer Quicksand*. Philadelphia: Altemus, 1912.

———. *The Young Engineers in Colorado; or At a Railroad Building in Earnest*. Philadelphia: Altemus, 1912.

———. *The Young Engineers in Mexico; or Fighting the Mine Swindlers*. Philadelphia: Altemus, 1913.

———. *The Young Engineers in Nevada; or, Seeking Fortune on the Turn of a Pick*. Philadelphia: Altemus, 1913.

Harner, John. "Place Identity and Copper Mining in Sonora, Mexico." *Annals of American Geographers* 91, no. 4 (2001):660–680.

Harpending, Asbury. *The Great Diamond Hoax and Other Stirring Incidents in the Life of Asbury Harpending*. [New York]: James H. Barry, Co, 1913.

Harvey, Charles, and Jon Press. "Overseas Investment and the Professional Advance of British Metal Mining Engineers, 1851–1914." *Economic History Review*, new series 42, no. 1 (1989):64–86.

Historical Statistics of the United States, Table DB73-78, Millennial Edition Online.

Hittell, John Shertzer. *Mining in the Pacific States of North America*. San Francisco: H. H. Bancroft, 1861.

Horsman, Reginald. *Race and Manifest Destiny: The Origins of American Racial Anglo-Saxonism*. Cambridge, MA: Harvard University Press, 1981.

Hounshell, David. *From the American System to Mass Production: The Development of Manufacturing Technology in the United States*. Baltimore, MD: Johns Hopkins, 1984.

D. Houston and Co., *Copper Manual: Copper Mines, Copper Statistics, Copper Shares*. New York: D. Houston, 1899.

Hovis, Logan, and Jeremy Mouat. "Miners, Engineers, and the Transformation of Work in the Western Mining Industry, 1880–1930." *Technology and Culture* 37, no. 3 (1996):429–456.

Hugginie, A. Yvette. "A New Hero Comes to Town: The Anglo Mining Engineer and 'Mexican Labor' as Contested Terrain in Southeastern Arizona, 1880–1920." *New Mexico Historical Review* 69, no. 4 (1994):323–344.

———. "'Strikitos!': Race, Class, and Work in the Arizona Copper Industry, 1870–1920." PhD diss., Yale University, 1991.

Hughes, Thomas P. *Networks of Power: Electrification in Western Society, 1880–1930*. Baltimore, MD: Johns Hopkins University Press, 1983.

Hyde, Charles K. *Copper for America: The United States Copper Industry from Colonial Times to the 1990s*. Tucson: University of Arizona Press, 1998.

Hyndman, Henry Mayers. *The Record of an Adventurous Life*. New York: MacMillan, 1911.

Ingersoll, Ralph McA. *In and Under Mexico*. New York: Century, 1924.

Isenberg, Andrew. *Mining California: An Ecological History*. New York: Hill and Wang, 2005.

Jacobsen, Matthew Frye. *Whiteness of a Different Color: European Immigrants and the Alchemy of Race*. Cambridge, MA: Harvard University Press, 1998.

Jacoby, Karl. *Shadows at Dawn: A Borderlands Massacre and the Violence of History*. New York: Penguin, 2008.

Jameson, Elizabeth. *All that Glitters: Class, Conflict, and Community in Cripple Creek*. Urbana: University of Illinois Press, 1998.

Johnson, Susan Lee. *Roaring Camp: The Social World of the California Gold Rush*. New York: Norton, 2000.

Jones, Chris. "A Landscape of Energy Abundance: Anthracite Coal Canals and the Roots of American Fossil Fuel Dependence, 1820–1860." *Environmental History* 15, no. 3 (2010):449–484.

Joralemon, *Romantic Copper: Its Lure and Lore*. New York: D. Appleton-Century, 1934.

Kingdon, Maud Kenyon. *From Out of Dark Shadows*. San Diego, CA: Frye and Smith, 1924.

Kline, Ronald R. *Consumers in the Country: Technology and Social Change in Rural America*. Baltimore, MD: Johns Hopkins University Press, 2000.

———. "From Progressivism to Engineering Studies: Edwin T. Layton's *Revolt of the Engineers* Reconsidered." *Technology and Culture* 49, no. 4 (2008):1018–1024.

Kranakis, Eda. "Social Determinants of Engineering Practice: A Comparative View of France and America in the Nineteenth Century." *Social Studies of Science* 19, no. 1 (1989):5–70.

Küstel, Guido. *Roasting of Gold and Silver Ores: And the Extraction of Their Respective Metals, Without Quicksilver*. San Francisco: Dewey, 1870.

———. *A Treatise on Concentration of All Kinds of Ores: Including the Chlorination Process for Gold-Bearing Sulphurets, Arseniurets, and Gold and Silver Ores Generally*. San Francisco: Mining and Scientific Press, 1868.

———. *Nevada and California Processes of Silver and Gold Extraction: For General Use, and Especially for the Mining Public of California and Nevada: With Full Explanations and Directions for All Metallurgical Operations Connected with Silver and Gold*. San Francisco: Frank D. Carlton, 1863.

LaFeber, Walter. *The New Empire: An Interpretation of American Expansion, 1860–1898*. 35th anniversary edition. Ithaca, NY: Cornell University Press, 1998.

Lamar, Howard. *The Far Southwest, 1846–1912: A Territorial History*. Revised edition. Albuquerque: University of New Mexico Press, 2000.

Lankton, Larry D., and Jack K. Martin. "Technological Advance, Organizational Structure, and Underground Fatalities in the Upper Michigan Copper Mines, 1860–1929." *Technology and Culture* 28, no. 1 (1989):42–66.

Lankton, Larry D. *Cradle to Grave: Life, Work, and Death at the Lake Superior Copper Mines*. New York: Oxford University Press, 1991.
Layton, Edwin. *The Revolt of the Engineers: Social Responsibility and the American Engineering Profession*. Cleveland, OH: Case Western Reserve University Press, 1971.
———. "Veblen and the Engineers." *American Quarterly* 14, no.1 (1962):64–72.
LeCain, Tim. *Mass Destruction: The Men and Giant Mines that Wired America and Scarred the Planet*. New Brunswick, NJ: Rutgers University Press, 2009.
Lewis, Daniel. *Iron Horse Imperialism: The Southern Pacific of Mexico, 1880–1951*. Tucson: University of Arizona Press, 2007.
Lingenfelter, Richard. *The Hardrock Miners: A History of the Mining Labor Movement in the American West, 1863–1893*. Berkeley: University of California Press, 1974.
Lounsbury, Thomas Raynesford. *Sheffield Scientific School, 1847–1879. An Historical Sketch*... [New Haven: Press of Tuttle Morehouse and Taylor, n.d.].
Marcosson, Isaac F. *Metal Magic: The Story of the American Smelting and Refining Company*. New York: Farrar, Strauss, 1949.
Marcus, Alan, ed. *Engineering in a Land-Grant Context: The Past, Present, and Future of an Idea*. West Lafayette, IN: Purdue University Press, 2005.
Martin, Brian, ed. *Confronting the Experts*. Albany: SUNY Press, 1993.
Marx, Leo. *The Machine in the Garden: Technology and the Pastoral Ideal in America*. New York: Oxford University Press, 1964.
May, Ernest. *The Southern Dream of a Caribbean Empire*. Baton Rouge: Louisiana State University Press, 1973.
McGivern, James Gregory. *First Hundred Years of Engineering Education in the United States (1807–1907)*. Spokane, WA: Gonzaga University Press, 1960.
Meiksin, Peter, and Chris Smith, eds. *Engineering Labor: Technical Workers in Comparative Perspective*. London: Verso, 1996.
Meinig, D. W. *The Shaping of America: A Geographical Perspective on 500 Years of History*. New Haven, CT: Yale University Press, 1998.
Morison, Samuel Eliot. *The Tercentennial History of Harvard College and University, 1636–1936*. Vol. 1, *The Development of Harvard University since the Inauguration of President Eliot 1869–1929*. Cambridge, MA: Harvard University Press, 1930.
Morse, Kathryn. *The Nature of Gold: An Environmental History of the Klondike Gold Rush*. Seattle: University of Washington Press, 2003.
Mouat, Jeremy, and Ian Phimister. "The Engineering of Herbert Hoover." *Pacific Historical Review* 77, no. 4 (2008):553–584.
Mowry, Sylvester. *Arizona and Sonora: The Geography, History, and Resources of the Silver Region of North America*. 3rd ed. New York: Harper and Bros., 1866.
Noble, David. *America by Design: Science, Technology, and the Rise of Corporate Capitalism*. New York: Knopf, 1977.
North, Diane M. T. *Samuel Peter Heintzelman and the Sonora Exploring and Mining Company*. Tucson: University of Arizona Press, 1980.
Nye, David. *Electrifying America: The Social Meanings of a New Technology*. Cambridge, MA: MIT Press, 1990.
Nystrom, Eric. *Seeing Underground: Maps, Models, and Mining Engineering in America*. Reno: University of Nevada Press, 2014.

Ochs, Kathleen. "The Rise of American Mining Engineers: A Case Study of the Colorado School of Mines." *Technology and Culture* 33, no. 2 (1992):278–301.

Oldenziel, Ruth. *Making Technology Masculine: Men, Women, and Modern Machines in America, 1870–1945*. Amsterdam, NL: Amsterdam University Press, 1999.

Osbourne, Henry Stafford. *A Practical Manual of Minerals, Mines, and Mining*. 2nd ed., rev. Philadelphia: Henry Carey Baird and Co., 1896.

Park, Joseph F. *The History of Mexican Labor in Arizona During the Territorial Period*. Master's thesis, University of Arizona, 1961.

Parrish, Michel E. *Mexican Workers, Progressives, and Copper: The Failure of Industrial Democracy during the Wilson Years*. La Jolla, CA: Chicano Research Publications, 1979.

Paul, Rodman Wilson. *Mining Frontiers of the Far West: 1848–1880*. Expanded edition, ed. Elliot West. Albuquerque: University of New Mexico Press, 2001.

———, ed. *A Victorian Gentlewoman in the Far West: The Reminiscences of Mary Halleck Foote*. San Marino, CA: The Huntington Library, 1972.

Peck, Gunther. "Manly Gambles: The Politics of Risk on the Comstock Lode, 1860–1880." *Journal of Social History* 26, no. 4 (1993):701–723.

Peele, Robert. *Mining Engineer's Handbook*, vol. 1. New York: Wiley, 1918.

Perales, Monica. *Smeltertown: Making and Remembering a Southwest Border Community*. Chapel Hill: University of North Carolina Press, 2010.

Peters, Edward Dyer. *Modern Copper Smelting*. 7th ed. New York: Scientific Publishing, 1895.

Peterson, Richard H. "The Frontier Thesis and Social Mobility on the Mining Frontier." *The Pacific Historical Review* 44, no. 1 (1975):52–67.

Poston, Charles. *Building a State in Apache Land: The Story of Arizona's Founding as Told by Arizona's Founder*. Edited by John Myers Myers. Tempe, AZ: Aztec Press, 1963.

Press Reference Library (Western Edition). *Being the Portraits and Biographies of the Progressive Men of the West*. Vol 2. New York: International News Service, 1915.

Pritchard, Sara. *Confluence: The Nature of Technology and the Remaking of the Rhône*. Cambridge, MA: Harvard University Press, 2011.

Pumpelly, Raphael. *Across America and Asia: Notes of a Five Years' Journey Around the World, and of Residence in Arizona, Japan, and China*. 2nd ed., rev. New York: Leypoldt and Holt, 1870.

———. *My Reminiscences*. New York: Holt, 1918.

Rae, John B. "Engineers Are People." *Technology and Culture* 16, no. 3 (1975):404–418.

Raymond, Rossiter. *Silver and Gold: An Account of the Metallurgical Industry of the United States, with Reference Chiefly to the Precious Metals*. New York: J. B. Ford, 1873.

———. *Statistics of Mines and Mining in the States and Territories*...Vol. 869. Washington, D.C.: Government Printing Office, 1870.

Read, Thomas Thornton. *The Development of Mineral Industry Education in the United States*. New York: American Institute of Mining and Metallurgical Engineers, 1941.

Reuss, Martin. "Seeing Like an Engineer: Water Projects and the Mediation of the Incommensurable." *Technology and Culture* 49, no. 3 (2008):531–546.

Rickard, T. A. *The Copper Mines of Lake Superior*. New York: Engineering and Mining Journal Press, 1905.

———. *A History of American Mining*. New York: McGraw Hill, 1932.

———. *Interviews with Mining Engineers*. San Francisco: Mining and Scientific Press, 1922.

Ritter, Etienne. *From Prospect to Mine*. Denver: Mining Science Publishing, 1910.

Robbins, William G. *Colony and Empire: The Capitalist Transformation of the American West*. Lawrence: Kansas University Press, 1994.

Roosevelt, Theodore. *The Works of Theodore Roosevelt*. New York: Century, 1901.

Rotundo, Anthony. *American Manhood: Transformations in American Masculinity from the Revolution to the Modern Era*. New York: Basic Books, 1993.

Sabin, Paul. "Home and Abroad: The Two 'Wests' of Twentieth-Century United States History." *Pacific Historical Review* 66, no. 3 (1997):305–335.

Sachs, Aaron. *The Humboldt Current: Nineteenth-Century Exploration and the Roots of American Environmentalism*. New York: Viking, 2006.

Salas, Miguel Tinker. *In the Shadow of the Eagles: Sonora and the Transformation of the Border during the Porfiriato*. Berkeley: University of California Press, 1997.

———. "Sonora: The Making of a Border Society, 1880–1910." *Journal of the Southwest* 34, no. 4 (1992):429–456.

Sandweiss, Martha A. *Passing Strange: A Gilded Age Tale of Love and Deception Across the Color Line*. New York: Penguin, 2009.

[Santa Rita Silver Mining Company]. *Second Annual Report of the Santa Rita Silver Mining Company, Made to the Stockholders, March 19, 1860*. Cincinnati, OH: Railroad Record Print, 1860.

Schmitz, Christopher. "The Rise of Big Business in the World Copper Industry, 1870–1930." *Economic History Review*, new series 39, no. 3 (1986):392–410.

Schwantes, Carlos A. *Vision and Enterprise: Exploring the History of Phelps Dodge Corporation*. Tucson: University of Arizona Press, 2000.

Scott, James. *Seeing Like a State: How Certain Schemes to Improve the Human Condition Have Failed*. New Haven, CT: Yale University Press, 1999.

Servos, John W. "Mathematics and the Physical Sciences in America, 1880–1930." *Isis* 77, no. 4 (1986):611–629.

Smith, Duane. *Mining America: The Industry and the Environment*. Boulder: University Press of Colorado, 1994.

———. *Rocky Mountain Mining Camps: The Urban Frontier*. Lincoln: University of Nebraska Press, 1973.

Smith, Merritt Roe and Leo Marx, eds. *Does Technology Drive History?* Cambridge, MA: MIT Press, 1994.

[Sonora Exploring and Mining Company]. *Sonora and the Value of Its Silver Mines: Report of the Sonora Exploring and Mining Co., Made to the Stockholders, December, 1856*. Cincinnati, OH: Railroad Record Print, 1856.

———. *Third Annual Report of the Sonora Exploring and Mining Co., Made to the Stockholders, March 1859*. New York: W. Minns, 1859.

———. *Fourth Annual Report*. New York: W. Minns, 1860.

Spence, Clark. *Mining Engineers and the American West: The Lace-Boot Brigade*. New Haven: Yale University Press, 1970.

Spude, Robert L., "Frank Morrill Murphy, 1854–1917: Mining and Railroad Mogul and Developer of the American Southwest." In *Mining Tycoons in the Age of Empire: Entrepreneurship, High Finance, Politics, and Territorial Expansion*, edited by Raymond E. Dumett1, 151–170. Burlington, VT: Ashgate, 2009.

Stegner, Wallace. *Angle of Repose*. Garden City, NY: Doubleday, 1971.

Stetefeldt, Carl August. *The Lixiviation of Silver-Ores with Hyposulphite Solutions, with Special Reference to the Russell Process*. New York: Scientific Publishing, 1888.

Steinberg, Ted. *Down to Earth: Nature's Role in American History*. New York: Oxford University Press, 2002.

Stevens, Horace J., ed. *The Copper Handbook*, vol. 3. Houghton, MI: Horace J. Stevens, 1903.

———, ed. *Mines Register: A Successor to the Mines Handbook and the Copper Handbook*. Vol. 3. Houghton, MI: Horace J. Stevens, 1902.

St. John, Rachel. *Line in the Sand: A History of the Western U.S.-Mexico Border*. Princeton, NJ: Princeton University Press, 2011.

Teisch, Jessica. *Engineering Nature: Water, Development, and the Global Spread of American Environmental Expertise*. Chapel Hill: University of North Carolina Press, 2011.

Trachtenberg, Alan. *The Incorporation of America: Culture and Society in the Gilded Age*. New York: Hill and Wang, 1982.

Truett, Samuel. *Fugitive Landscapes: The Forgotten History of the U.S.-Mexico Borderlands*. New Haven, CT: Yale University Press, 2008.

Turner, Frederick Jackson. *The Frontier in American History*. New York: Holt, 1921.

Turrell, Robert Vicat, with Jean-Jacques Van Helten. "The Rothschilds, the Exploration Company, and Mining Finance." *Business History* 28, no. 2 (1986):181–205.

Unger, Irwin, and Debi Unger. *The Guggenheims: A Family History*. New York: Harper, 2005.

Varady, Robert G., Helen Ingram, and Lenard Milich. "The Sonoran Pimería Alta: Shared Environmental Problems and Challenges." *Journal of the Southwest* 37, no. 1 (1995):102–122.

Veblen, Thorstein. *Engineers and the Price System*. New York: B. W. Huebsch, 1921.

Vitalis, Robert. *America's Kingdom: Mythmaking on the Saudi Oil Frontier*. Stanford, CA: Stanford University Press, 2007.

Voss, Stuart F. *On the Periphery of Nineteenth-Century Mexico: Sonora and Sinaloa, 1810–1877*. Tucson: University of Arizona Press, 1982.

Wasserman, Mark. *Capitalists, Caciques, and Revolution: The Native Elite and Foreign Enterprise in Chihuahua, 1854–1911*. Chapel Hill: University of North Carolina Press, 1984.

West, Elliott. *The Contested Plains: Indians, Goldseekers, and the Rush to Colorado*. Lawrence: University Press of Kansas, 1998.

White, Richard. *The Middle Ground: Indians, Empires, and Republics in the Great Lakes Region, 1650–1815*. New York: Cambridge University Press, 1991.

———. *Railroaded: The Transcontinentals and the Making of Modern America*. New York: Norton, 2011.

Wiebe, Robert H. *The Search for Order, 1877–1920*. New York: Hill and Wang, 1967.

Williams, Albert, Jr. *Mineral Resources of the United States, 1883–1884*. Washington, D.C.: Government Printing Office, 1885.

Williams, William Appleman. *The Tragedy of American Diplomacy*. New York: Norton, 1972.

Wyman, Mark. *Hard Rock Epic: Western Miners and the Industrial Revolution, 1860–1910*. Berkeley: University of California Press, 1989.

Young, Otis. *Western Mining: An Informal Account of Precious-Metals Prospecting, Placering, Lode Mining, and Milling on the American Frontier from Spanish Times to 1893*. Norman: University of Oklahoma Press, 1970.

INDEX

ABOUT THE AUTHOR

SARAH GROSSMAN is the editor of the Southeast Asia Program Publications (SEAP) imprint, and managing editor of the journal *Indonesia*, at Cornell University Press. She received her Ph.D. in U.S. history from the University of New Mexico.